Sustainable Practices
in the Built Environment

Sustainable Practices in the Built Environment

Second edition

Edited by
Craig A. Langston and Grace K.C. Ding

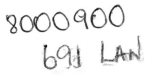

BUTTERWORTH HEINEMANN

OXFORD AUCKLAND BOSTON JOHANNESBURG MELBOURNE NEW DELHI

Butterworth-Heinemann
Linacre House, Jordan Hill, Oxford OX2 8DP
225 Wildwood Avenue, Woburn, MA 01801-2041

ℛ A member of the Reed Elsevier plc group

First published 1997
Second edition 2001

British Library Cataloguing in Publication Data
Sustainable practices in the built environment
   1. Construction industry – Environmental aspects 2. Building materials –
   Environmental aspects
   I. Langton, Craig A.   II. Ding, Grace K. C.
   691

Library of Congress Cataloguing in Publication Data
Sustainable practices in the built environment/edited by Craig A. Langston and
Grace K.C.Ding
   p. cm
   Includes bibliographical references and index.
   ISBN 0-7506-5153-9
   1. Environmental engineering.   I. Langton, Craig A.   II. Ding, Grace K.C.
   TA170 .S88 2001
   628--dc21

                          2001035007

ISBN 0 7506 5153 9

Typeset by Avocet Typeset, Brill, Aylesbury, Bucks
Printed and bound in Great Britain

Contents

List of contributors   vii
Foreword to first edition   ix
Foreword to second edition   xi
Introduction   xiii

PART 1: ENVIRONMENTAL QUALITY   **1**
1. The planet in crisis   3
2. Sustainable development   15
3. Accounting for the environment   25

PART 2: DEVELOPMENT CONTROLS   **35**
4. Environmental law   37
5. Environmental impact assessment   44
6. Environmental policies and strategies   51

PART 3: ANALYTICAL TOOLS   **61**
7. Environmental economics   65
8. Cost-benefit analysis   74
9. Estimating social costs and benefits   84

PART 4: PROJECT FEASIBILITY   **95**
10. Project selection criteria   97
11. Intergenerational equity   107
12. The measurement of sustainability   116

PART 5: DESIGN CONSIDERATIONS   **125**
13. Environmental impact of buildings   127
14. Low energy design   136
15. Embodied energy and recycling   150

PART 6: ENERGY CONSERVATION   **163**
16. Energy quality   167

17. Renewable energy   176
18. Energy regulation and policy   193

PART 7: LIFE-COST STUDIES   **203**
19. Life-cost planning and analysis   205
20. Determination of the discount rate   218
21. Occupancy costs   230

PART 8: ASSET MANAGEMENT   **249**
22. Post occupancy evaluation   253
23. Environmental auditing   261
24. Facility management   270

Index   283

Contributors

Editors

Craig Langston – Associate Professor in Construction Economics at the University of Technology, Sydney (UTS). Before commencing at UTS, he worked for a professional quantity surveying office in Sydney. His PhD thesis was concerned with discounting and life-cost studies. He developed two cost-planning software packages (PROPHET and LIFECOST) that are sold internationally, and is author of several books covering aspects of construction economics and facility management.

Grace Ding – Lecturer in Construction Economics at UTS. She has a diploma from Hong Kong Polytechnic, a bachelors degree from the University of Ulster and a masters degree by thesis from the University of Salford. She has practised as a quantity surveyor in Hong Kong, England and Australia. Grace is currently undertaking PhD research in the area of environmental performance measurement.

Specialist contributions

Rick Best – Lecturer in Construction Economics at UTS. He has degrees in architecture and quantity surveying and has research interests in information technology, energy in buildings and low energy design strategies. Rick is undertaking research in district energy systems and the international performance of construction projects, and is the co-editor of several books on value in building.

Gerard de Valence – Senior Lecturer in Construction Economics at UTS. He has an honours degree in economics from the University of Sydney. He has worked in industry as an analyst and economist. His principal areas of research interest include the measurement of project performance, economic factors related to the construction industry, and the impact of emerging technologies.

Jack Greenland – Retired academic. He is still one of the most respected building scientists in Australia and has lectured to architects and other professionals for over 25 years. He is author of a leading text on heat, light and sound and continues to undertake research and PhD supervision in a range of areas related to energy in buildings.

Rima Lauge-Kirstensen – Architect and research scholar. She holds a PhD and a bachelors degree in architecture, and has worked in a number of roles relating to the construction industry. She has particular expertise in the physics of building systems and emerging technologies that aim to minimize energy demand.

Caroline Mackley – Project Manager and Researcher with Bovis Lend Lease. She holds a bachelors degree in quantity surveying, a masters degree in environmental studies and is currently undertaking a PhD in the area of embodied energy. She has taught at Massey University in New Zealand and has conducted her own practice in Australia.

Peter Smith – Senior Lecturer in Construction Economics at UTS. Prior to his current appointment he worked for a large professional quantity surveying practice and an international construction and property development company. He is currently undertaking PhD research and has an interest in consumer investment advice and property maintenance.

Other contributors

The valuable contribution of the following people to the writing of this book is also acknowledged and greatly appreciated:

Yu Lay Langston
Ann Godfrey
Phyllis Campbell
Paul Pholeros
Anne-Marie Willis
Alan Gilpin
Margaret Durham
Chris Hall
Martin Hill
Ian Wills
Merv Fiedler
David Muir
Roger Horwood
Bill Lawson
Robert Turner
Stephen Ballesty

Foreword
to first edition

Allan Ashworth

'We know that the whole creation has been groaning as in the pains of childbirth right up to the present time.' The Book of Romans, Chapter 8, Verse 22.

The importance of environmental issues throughout society has, I am pleased to say at last, raised its profile. This increased emphasis is likely to be continued both for our benefit and those of future generations. It is also becoming much more recognized as a subject of study in universities and colleges around the world. Increasingly the activities of the construction industry through mineral extraction, materials and component manufacture, building and infrastructure development and through the use of buildings and structures has concentrated the attention of designers and constructors on the environment. More action and greater care are required to be taken. Thus those involved in the design and construction process and others concerned in the important area of providing better living, working and leisure conditions are encouraged to respond to alternative and improved accommodation solutions, taking into account the needs of individual projects. Such solutions must now more fully consider the environmental implications of development.

This book, written by experts in the field of environmental economics, presents its readers not only with the scenarios to be addressed, but solutions to how development can be undertaken in an environmentally efficient way. The book does not limit itself to the theory alone, but provides a basis for real alternatives in respect of ecologically sustainable development and addresses the economic approaches and implications that bear on the construction industry. The authors have deliberately dealt with the general, practical and detailed in such a way as to offer a robust analysis, reminding the reader that the world's resources are not infinite.

It is important for all those who want to see a prosperous construction industry to concern themselves with the issues of ecologically sustainable development. Such a concern must not be limited to the location of the project alone, but the broader impacts of such development must also be taken into account.

Environmental impact assessment has now become an important consideration in the developed world. Such analyses must not be allowed to be restricted to 'one's own backyard' but they must also consider the wider impact of environmental issues that such development raises in a global context.

I recommend this book to all those people who share my concern about the environment, the prosperity of future generations and the responsibility entrusted with those involved in the construction industry to properly consider the impacts of their decisions.

Allan Ashworth HMI Msc ARICS
Chartered Quantity Surveyor
York, England
1996

Foreword to second edition

Dr Douglas Ferry

Since this book was first conceived things have moved on.

Although designers were becoming increasingly aware of long-term environmental issues during the last part of the twentieth century, short-term profitability for the developer was still tending to win out. Environmental responsibility and the long-term view were seen as a somewhat idealistic approach suitable for those Mr Nice Guys who were able to afford it.

After all, it was argued, what effect could a single building have upon a multi-billion pound national construction programme, or upon the astronomical figures of global construction resource-use?

It is worth noting that, even on its own terms, such an escapist philosophy ignored the fact that many of the most pressing environmental issues are local, where a single development can have a considerable effect for good or bad.

But today government initiatives in the form of penalties and incentives mean that even those developers who refuse to see the probable benefit to themselves in taking a long-term view are forced to consider the environmental consequences of what they are doing.

And here a hopeful comparison is seen with automobile design, where for many years safety considerations were thought to be unattractive from a marketing viewpoint. But once legislation forced attention on this by the compulsory use of seat-belts, consumers became increasingly safety conscious and such things as airbags, side-impact protection, and ABS braking became the norm without the need for government action.

So all those concerned with building procurement need to be properly informed about sustainable construction, and this book is a veritable mine of information, progressing from clear explanations of the strategic issues into methods of implementation. Each chapter deals comprehensively with its subject in a hard-headed real-world context, and includes an exhaustive list of references and further reading.

Douglas J. O. Ferry PhD FRICS
Adjunct Professor UTS
2001

Introduction

We live in an environmental age. As the environmental crisis deepens, valuable resources are further depleted and limits to growth are approached, it is even more critical that development of built infrastructure takes proper cognizance of environmental impact.

This book is about making development and construction more sustainable. It is written specifically for people working in or studying about the construction industry. It embraces a series of topics that address sustainable practices: the key issues, implementation strategies, policy initiatives and case studies. The aim of this book is to show ways in which people involved in and allied to the industry can make a difference by delivering projects that reflect sustainability goals. It, however, does not deal with the broader issues of urban planning and transportation, which are important topics in their own right but outside the intended scope of this text. In the next decade the construction industry will be judged by its success in this endeavour, and increasing levels of government intervention can be expected if it fails to live up to its responsibilities.

The philosophy of sustainable development borrows freely from the science of environmental economics in several major respects. A basic component of environmental economics concerns the way in which economics and the environment interact. Fundamental to an understanding of sustainable development is the fact that the economy is not separate from the environment. There is an interdependence because the way humans manage the economy impacts on the environment and the resultant environmental quality, in turn, impacts on the future performance of the economy.

Definitions of sustainable development abound since what constitutes development for one person may not be development or progress for another. Development is essentially a value word: it embodies personal aspirations, ideals and concepts of what comprises a benefit for society. The most popular definition of sustainable development is the one given in the Brundtland Report (1987):

> development that meets the needs of the present without compromising the ability of future generations to meet their own needs.

This definition is about the present generation's stewardship of resources. It means that for an economic activity to be sustainable it must neither degrade nor deplete the natural resources nor have serious impacts on the global environment inherited by future generations. For example, if greenhouse gases continue to build, the ozone layer becomes further depleted, soil quality is degraded, natural resources are exhausted and water and air are badly polluted, then the present generation clearly has prejudiced the ability of future generations to support themselves.

Development implies change, and should by definition lead to an improvement in the quality of life of individuals. Development encompasses not only growth, but also general utility and well-being, and involves the transformation of natural resources into productive output. Sustainable development in practice represents a balance (or compromise) between economic progress and environmental conservation in much the same way as value for money on construction projects is a balance between maximum functionality and minimum (life-)cost. The economy and the environment necessarily interact, and so it is not appropriate to focus on one and ignore the other.

Development is undeniably associated with construction and the built environment. Natural resources are consumed by the modification of land, the manufacture of materials and systems, the construction process, energy requirements and the waste products that result from operation, occupation and renewal. Building projects are a major contributor to both economic growth and environmental degradation and hence are intimately concerned with sustainable development concepts. Every project (new or existing) can be enhanced by consideration of 'whole-of-life' methodologies. The term 'whole-of-life' simply means taking a long-term view and finding a balance between the various attributes of a project, including how to properly evaluate the impact of present and future benefits.

The concept of sustainable development has stimulated the search for construction solutions that do not result in a clash between rising living standards and environmental protection. The essential ingredients for sustainability seem clear, although how they are brought together in practice is undeniably complex. The main challenges are:

1. The construction industry must re-engineer its entire production process.
2. Current economic approaches to project evaluation need to be re-examined.
3. The integration of economic, environmental and social aspects of sustainability needs to be further developed.
4. There needs to be increased awareness and understanding of sustainability issues at all levels in the community.

The chapters in this book are categorized in a matrix arrangement to illustrate their interrelationship. The order of chapters reflects a progression from macro issues to micro issues. The vertical and horizontal integration of the chapters is shown in the diagram below.

Vertical integration is used to differentiate macro and micro issues. Chapters 1 to 12 cover strategic planning topics, while Chapters 13 to 24 cover facility design and management topics. Horizontal integration is used to differentiate environmental and economic issues. Chapters 1 to 6 and 13 to 18 cover topics about

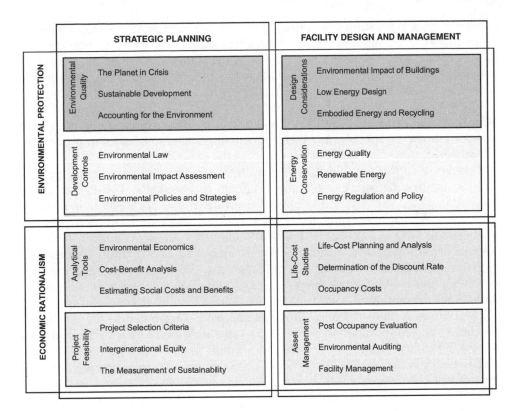

environmental protection, while Chapters 7 to 12 and 19 to 24 cover topics about economic rationalism. Furthermore, chapters are grouped into eight parts and categorized by common themes.

Strategic planning concerns those aspects of sustainable development that are about making effective decisions between alternative courses of action. Topic areas include environmental quality, development controls, analytical tools and project feasibility, and are relevant to the pre-design stage of the development process. Arguably this is the most important stage, although to some extent it is also the most intangible. The first half of the book concludes with recommendations on how sustainable development can be objectively measured and used to make appropriate decisions.

Facility design and management concerns those aspects of sustainable development that relate to specific project choices. Topic areas include design considerations, energy conservation, life-cost studies and asset management and are relevant to design, construction and occupation stages of the development process. The second half of the book concludes by suggesting that the effective management of facilities, which typically spans long time horizons, is critical to the realization of benefits not only to the owner but to society as a whole.

The chapters categorized as environmental protection involve a study of the environmental side of the sustainability equation. While there may be a tendency to maximize the conservation of resources, the reality is that a balance between conservation and consumption is necessary. The chapters categorized as economic rationalism present the methods for assessment and management of development, and therefore focus on the other side of the sustainability equation. Monetary issues cannot be ignored, but instead must be integrated into development processes so that balanced decisions can be made.

The strength of this book is in its ability to present a balanced argument and to offer solutions that are both effective and practical. The chapters are written from an international perspective and therefore are applicable to readers from any part of the world.

This book forms a useful introductory text for construction industry professionals, facility managers, construction clients and students. The layout of the book reflects the structure of the *Environmental Economics* subject in the Master of Facility Management and the Master of Business Administration (Facility Management) degrees at the University of Technology, Sydney. Graduates of these courses are employed in a range of fields including engineering, architecture, property management, construction management, project management, quantity surveying and facility management, and apply their skills to projects in countries throughout the world.

The editors wish to thank all those people who contributed to the writing of this book, and to those who gave up their valuable time to provide input and/or review individual chapters. The publishers would like to thank Dr Jorge Vanegas of The Georgia Institute of Technology, USA, for his help in reviewing the final manuscript.

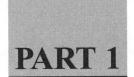

PART 1

Environmental quality

The environment and the economy can no longer be treated as independent and the subject of separate policy direction. Rather, the environment supports economic growth and rising living standards, providing vital social amenity that itself encourages human productivity and creativity. Economic well-being supports initiatives that can protect, conserve and improve environmental quality, fund research and assist in new discoveries or better understanding of natural phenomena.

This balance between environmental conservation, which aims to protect natural resources, and economic progress, which aims to develop human infrastructure, is what is known as sustainable development. Too much emphasis on the environmental side will limit the ability to deliver improvements in living standards, particularly for the developing world, while too much emphasis on the economic side will lead to depletion of vital natural resources that cannot be readily recreated.

Therefore sustainable development is one of the most important issues facing the quality of human habitation of the planet in the future, and one which must be addressed now if an effective balance is to be struck. It involves proper valuation of environmental goods and services, taking a long-term view of development decisions, and seeking to provide equity both within the current generation (rich versus poor) and across generations (present versus future).

The construction industry is a major player in arriving at an effective balance, as most new projects involve some form of resource consumption and site modification that, as a general rule, diminishes environmental wealth and increases capital wealth. So projects which minimize impact on the environment while still providing necessary economic and social advantage should be favoured. Such an approach must bring together the two sides of the equation during the decision-making process. Techniques are needed to assist in this endeavour so that a concentration on short-term monetary performance does not dominate.

In fact, a preoccupation with monetary evaluation has worked against the implementation of sustainable practices. Environmental impact and financial return have been separately considered, but seldom integrated into a single decision criterion. This has led to a division of subjective and objective performance measures; the former influenced by political interpretations of societal need and the latter influenced by business goals and profit maximization. Not until subjective and objective criteria can be evaluated together will sustainable development goals ever be realistically approached.

But in reality there is probably no such thing as sustainable development, at least for the majority of new projects that are commenced every day. The same can be said for existing infrastructure, which may have had the added disadvantage of poor environmental design in the first place. Sustainable development is not a point we reach but a journey we take. It is therefore an ideal, a set of goals, an objective to be pursued, but seldom realized. Yet that is not to say that the concept is a waste of effort. On the contrary, every step towards it is a positive contribution. While most projects will consume more resources than they create, projects that are closer to sustainable ideals will increasingly deliver benefits to their owners and users and to society as a whole. Therefore if design can encompass assessment and decision-making processes that address sustainability goals, it is likely that over the long term the construction industry will be able to demonstrate a significant contribution to global resource efficiency.

Many of the major environmental issues that society faces, however, are not related to individual projects or decisions, but rather are the result of vast collections of projects or decisions on a global basis. So how can any one project, or any one country, make a difference to environmental quality? The answer is reflected in the current debate over recycling. It is true that no one person can make a difference by recycling waste products from their own activities, but if everyone does it then a significant benefit can be realized. In the same way, if every project meets strict criteria for design, construction and operation then over a period of time problems like greenhouse gas emissions, global warming, pollution and loss of biodiversity can be turned around. These are international problems but can be overcome by local solutions. Everything is a piece of a much bigger puzzle, so a concerted effort across the board is required.

Yet exponential increases in population growth are the most serious long-term threat, with the capacity to adversely influence all other environmental problems by placing stress on natural systems to support and assimilate activity. The construction industry, in providing additional infrastructure for a growing population, will have even further impact on environmental goods and services, and as such makes an even stronger case to search for ways to make development more sustainable.

The chapters in this part deal with the important area of environmental quality. Chapter 1 discusses the crisis that afflicts the environment and shows the connectivity and complexity of key global problems. Chapter 2 explores the concept of sustainable development and its characteristics. Chapter 3 demonstrates that environmental goods and services must be properly valued and incorporated into economic models so that environmental wealth and capital wealth can be assessed collectively.

Environmental quality is both the start and the end of the sustainable development debate. It is both the rationale and the objective; the problem and the solution. Only through an appreciation of the importance of the environment to human prosperity can development be put into its proper perspective.

1

The planet in crisis

Caroline Mackley

1.1 Introduction

It is commonly accepted that the planet faces an environmental crisis precipitated by anthropocentric activity that is resulting in a reduction in the earth's productive capacity from which serious consequential social and environmental effects are starting to flow. The significance of the problem has given rise to global co-operation in the form of a range of major international agreements constructed with the objective of seeking a balance between the opposing yet interdependent forces of society, economy and environment.

Nowhere is the complexity and importance of this relationship more evident than in relation to the built environment. Constructed facilities are humankind's most important economic, social and environmental investment. When viewed in terms of its economic significance (as measured as a proportion of GDP), the direct and indirect capital flows constitute on average about 40% of national GDP (Bon and Pietroforte, 1999; International Building Research Council, 1999). In addition, 50% of the world's primary energy production is consumed by buildings when both direct and indirect flows are considered together (Levine et al., 1995; Russell, 1998).

It has been noted that our current economic activity has the potential to reduce the capacity of the environment to provide useful inputs to, and services for, future economic activity (Dovers, 1994). Human economic activity is the principal cause of the environmental crisis through exploitation and pollution, and yet such activity relies on a healthy environment for its continuance and productivity. There is hence a vital partnership upon which much depends.

Sustainability relates to the carrying capacity of the planet to maintain life in all its forms. It is often referred to as 'sustainable harvests', or in other words, the amount that various resources can be harvested at certain rates indefinitely without decreasing environmental quality or necessitating reductions in carrying capacity. At a global level sustainability is a complex issue and is difficult to determine accurately for any resource (Botkin and Keller, 1995).

Environmental quality and sustainability are closely linked. Future reductions in the quality and availability of natural resources upset estimates of carrying capacity and accelerate the likelihood of significant environmental catastrophe.

The environmental crisis concerns a wide range of matters. The major themes into which concerns are categorized are (a) social and economic dimensions, and (b) conservation and management of the natural environment. Social and economic dimensions concern such issues as poverty, consumption patterns, human population and health. Conservation and management issues cover atmospheric protection, land resources, ecosystem protection and waste management, amongst other things. The three main environmental problems currently facing the planet, however, are climate change, loss of biological diversity and population growth. These problems, as will be shown below, are interrelated.

1.2 Climate change

1.2.1 Atmospheric protection

'Of all the global environmental problems, climate change is the most pervasively threatening to human well-being and in many respects the most intractable' (Schipper and Meyers, 1994, p. 21).

The imperative to protect the atmosphere rises out of its various properties and the relationship it has with the maintenance and support of the Earth's ecosystems. It is now generally accepted that a link exists between atmospheric degradation and the onset of climate change and climate variability. Major uncertainties in modern scientific understanding of the causes and effects of climate change remain and thus are the focus of international research efforts. Whilst these uncertainties remain, the ability of governments to ratify increasingly demanding control measures is hindered.

Currently, the major concern in relation to climate change is anthropocentric air pollution in the form of carbon dioxide (CO_2), nitrogen oxides (NO_x), sulphur dioxide (SO_2), methyl chloride and chlorine gas (Cl_2) and methane (CH_4), as well as issues relating to ozone layer depletion. It is a major objective of the United Nations Conference for Environment and Development (UNCED) to 'improve the understanding of processes that influence and are influenced by the Earth's atmosphere on a global, regional and local scale including, inter alia, physical, chemical, geological, biological, oceanic, hydrological, economic and social processes ... to improve understanding of the economic and social consequences of atmospheric changes' (UNCED, 1996, p. 2).

In terms of quantifying impacts, atmospheric emissions are generally classified by impact categories instead of pure emissions of CO_2, SO_2, NO_x, etc. Each of these emissions may have impacts under several categories. These categories include global warming, depletion of the stratospheric ozone layer, acidification, nutrification, ecological toxicity and human toxicity (Fossdal, 1999).

1.2.2 The greenhouse effect

It is important to distinguish between the natural and enhanced greenhouse effect. Greenhouse gases play a vital role in maintaining a balance on Earth that keeps the temperature and functioning of ecosystems at a level conducive to human, animal and plant life – a delicate interaction that is not evident on any other planet in the solar system. The Earth's natural ability to balance the chemical composition of the atmosphere to maintain life is what is considered to be the natural greenhouse effect. The chemical composition of the atmosphere enables solar radiation to be absorbed at such a rate as to provide suitable temperature ranges and weather. It is estimated that the greenhouse balance, driven predominantly by carbon dioxide and atmospheric water vapour, leads to a mean surface temperature of the Earth of about 10°C above what it would otherwise be (Lovelock, 1979). Natural fluxes in this balance occur and up until the time of the industrial revolution, the Earth managed these without interference.

It is essential to be cognizant of the fact that greenhouse gases occur naturally in the atmosphere. Table 1.1 provides a summary of the composition of the Earth's atmosphere.

Table 1.1 Composition of the Earth's atmosphere

Gas	Proportion of atmosphere (%)	Flux in megatons per annum	Atmospheric function
Nitrogen	79	300	Pressure builder, fire extinguisher, alternative to nitrate in the sea
Oxygen	21	100 000	Energy reference gas
Carbon dioxide	0.03	140 000	Photosynthesis, climate control
Methane	10^{-4}	1 000	Oxygen regulation and ventilation of the anaerobic zone
Nitrous oxide	10^{-5}	30	Oxygen and ozone regulation
Ammonia	10^{-6}	300	PH control and climate control
Sulphur gases	10^{-8}	100	Transport gases of the sulphur cycle
Methyl chloride	10^{-7}	10	Ozone regulation

(*Adapted from: Lovelock, 1979, p. 68.*)

The enhanced greenhouse effect is caused by the build-up of greenhouse gases beyond known maximum natural atmospheric concentrations. This is compounded by the Earth's reduced capacity to absorb these as a result of the destruction of natural sinks.

Sources of greenhouse gas emissions are well known. Progress has been made in the last two decades towards a reduction in their atmospheric concentrations through a series of mitigation policies implemented as a result of major international agreements, but it should be noted that even if all emissions were to stop today, the atmospheric concentration of many of these substances would continue to increase for decades to come.

1.2.3 Global warming

Global warming is caused by the build-up of greenhouse gases like carbon dioxide, water vapour, methane, chlorofluorocarbons (CFCs), nitrous oxide and ozone which trap energy on the Earth's surface. While the presence of greenhouse gases is essential to the survival of all living creatures, increases upset the natural equilibrium. Scientists have measured a significant rise in the levels of heat-absorbing gases in the atmosphere and these increases give rise to global warming which can result in damaging consequences for the planet.

Historical evidence suggests that discernible climate change occurs in connection with a doubling or halving of atmospheric CO_2 concentrations. At present only about 0.03% of our atmosphere is carbon dioxide (see Table 1.1 above), which relates to approximately 6 gigatonnes of carbon dioxide emission per year (IPCC, 1995). During the most recent ice age 150 000 years ago, it has been calculated that the mean surface temperature of Earth was a mere 5°C cooler than today with corresponding CO_2 concentrations of 170 parts per million (ppm). In the interglacial period, about 100 000 years ago, the mean surface temperature was about 1 to 2°C warmer than today with concentrations of CO_2 of between 270 to 300 ppm (Falk and Brownlow, 1989). Since the time of the industrial revolution (about 1750 AD), atmospheric concentrations of CO_2 have increased from about 280 ppm to 360 ppm at present day (IPCC, 1995). This increase has corresponded with a rise in the global mean surface temperature of between 0.3 and 0.6°C since the late nineteenth century.

Climate change scenarios generally attempt to calculate the time frame and temperature change on the basis of a doubling of atmospheric carbon dioxide. By most accounts these scenarios point to the year 2100 as being the point at which significant changes will be clearly evident. The reason for international interest in reducing CO_2 emissions relates to the expected reduction in the Earth's productive capacity and the consequential social and environmental effects. The International Framework Convention on Climate Change (IFCCC) provides a protocol for all nations to measure and report their annual CO_2 emissions.

Reporting on the conclusions of various climate change studies, Leary and Scheraga (1994, p. 105) note that 'a 3°C warming will reduce annual world output by 1.3%'. These estimates are qualified for various assumptions and it is possible that a change in the basis of any variable could result in a sharp increase in this estimate. Global surface warming affects the climate pattern. Higher temperatures tend to speed up evaporation in some regions and cause more precipitation in others. A warmer atmosphere retains more moisture and water vapour absorbs more radiant heat.

The extent of knowledge supported by contemporary climate research suggests that the warming of the Earth's surface would have a significant effect on living creatures as well as the composition of the atmosphere. It is thought that oceans may expand and the polar ice caps may reduce in size which in turn will raise sea levels and release more carbon dioxide into the atmosphere, as oceans and ice caps are the planet's major carbon sinks. It is estimated that there will be an average increase in sea level of about 6 cm per decade for a temperature rise of between 1.5 to 5.5°C (Falk and Brownlow, 1989). The consequences of rising sea levels may be soil erosion, flooding and storm damage to many coastal regions as well as resultant economic and social losses.

1.2.4 Sources of greenhouse gas emissions

The major sources of greenhouse gas emissions originate from energy production and consumption, industrial developments, and land and marine resource use and development. Energy related CO_2 emissions account for 78% of global anthropogenic emissions (Schipper and Meyers, 1994) and therefore it is logical that the focus of assessment and mitigation continues to be linked to this aspect. Table 1.2 indicates greenhouse gas impacts and proportional source contributions.

Table 1.2 Greenhouse gas contributions

Indicator of impact	Baseline contribution	Energy production	Natural resource use	Industrial production
Land use or conversion	135 mil km²	6.2% fuelwood supply (6%), area occupied by energy facility (0.2%)	88% grazing, cultivation, cumulative desertification	6% lumbering towns, transport systems
Nitrogen fixation (as NO_x, NH_4)	200 mil tonnes/yr	32% fossil and traditional fuel burning	67% fertilizer, agricultural burning	1% refuse burning
Nitrous oxide emissions	7 mil tonnes/yr	16% fossil and traditional fuel burning	84% fertilizer, land clearing and aquifer disruption	Small
Carbon dioxide stock in atmosphere	280 ppm	78% fossil fuel burning	15% net deforestation for land clearing	7% net deforestation for lumber, cement manufacture
CO_2 particulate emissions to atmosphere	500 mil tonnes/yr	45% burning fossil and traditional fuels	40% biomass burning and wheat handling	15% smelting, non-agricultural land clearing
Methane stock in atmosphere	800 ppb	23% fossil fuel harvesting and processing	62% rice paddies, domestic animals and land clearing	15% landfills

(*Source: Schipper and Meyers, 1994, p. 18.*)

Carbon dioxide is mainly produced by the burning of compounds that contain carbon, such as coal, oil, gas and wood. In the past century, carbon dioxide levels have increased by 25% due to the burning of fossil fuels and the destruction of forests (Bates, 1990). Deforestation compounds carbon emissions as the carbon stored in the soil and in the trees and plants are returned to the air as CO_2.

Methane (CH_4) emissions are predicted to rise as natural and synthetic gases are used as alternatives to more expensive oil fuels. While gas is more efficient per unit of fuel than other forms of carbon fuels, it presents a difficult dilemma in relation to global warming. Although it is lower in concentration in the atmosphere than carbon dioxide, it has powerful heat-absorbing characteristics and each molecule has 30 or 40 times the warming potential of molecule of carbon dioxide. The increasing amount of methane in the air is primarily from the venting of oil and gas wells, but also is a result of conversion of forests and fields to cattle production and rice paddies, the harvesting of the oceans for fish, and the decay of organic wastes. Scientists believe that if current trends continue there will be at least twice as much methane in the air by the middle of the twenty-first century (New South Wales Government, 1990).

CFCs are compounds that contain carbon, fluorine, chlorine and sometimes hydrogen and are even more problematic as each molecule provides about 20 000 times more warming capacity than carbon dioxide (New South Wales Government, 1990). CFCs are exclusively man-made products used primarily in refrigerants, aerosol sprays, insulating materials and solvents. The presence of CFCs in the air not only traps heat but also attacks the stratospheric ozone layer.

Nitrous oxide (N_2O) accounts for about 5% of the greenhouse effect and the heat-absorbing power is about 200 times stronger than carbon dioxide. Nitrous oxide is mainly emitted by burning coal and oil, the denitrification of fertilizer and deforestation.

1.2.5 Depletion of the stratospheric ozone layer

Ozone is a layer of pale blue gas located in the atmospheric zone known as the stratosphere and occurring in a band between approximately 15 km to 50 km thick above the Earth's surface. Its essential function is to screen out damaging ultraviolet radiation, known as UV-B, from the sun. Ozone is a naturally occurring but rare form of oxygen. Ironically, ozone is kept in equilibrium (approx. 5 parts per million) by ultraviolet radiation (Lovelock, 1979).

Stratospheric ozone forms a protective layer that absorbs UV-B and therefore plays an important role in screening out high energy ultraviolet radiation from the sun. This radiation interferes with the human immune system, causes sunburn and contributes to certain kinds of skin cancers. It also reduces crop yield, and may even kill surface-dwelling fish and tiny marine organisms called phytoplankton. Some reports suggest that a major die off of the essential phytoplankton is plausible which may lead to a collapse in ocean ecosystems.

Ozone contains three oxygen atoms that break apart by ultraviolet photons. Through the natural process of ozone formation and destruction, most of the harmful and high energy radiation from the sun is screened before it reaches the Earth's surface. As a result of human activities the ozone equilibrium has been violated and new substances such as chlorine have been introduced into the stratosphere.

Over recent decades scientists have become increasingly concerned about the diminishing level of ozone in the stratosphere. The first clear sign of damage to the ozone layer was reported in 1985 by the British Antarctic Survey team who had been measuring ozone levels over the Antarctic since 1957. In October 1987, when this hole was very severe, the total amount of ozone measured was less than half of its 1970 levels. In 1993 the Antarctic ozone hole was the deepest ever recorded, with two thirds of the protective ozone shield in this area destroyed (Gribbin, 1988). Since then a hole over the Arctic region has also been discovered.

The instruments in place which seek to preserve the stratospheric ozone layer include the 1985 Vienna Convention for the Protection of the Ozone Layer, the 1987 Montreal Protocol on Substances that Deplete the Ozone Layer and the 1992 United Nations Framework Convention on Climate Change. Up until the Montreal Portocol was ratified, CFCs were widely used as coolants, propellants and solvents globally. Soon after the agreement, participating countries further agreed to phase out all CFC production by the year 2000. Other countries, notably China, also became signatories. However the success of the protocol is yet to be fully determined. It is up to each country to monitor and report on its success. No enforcement provisions have ever been agreed.

Apart from the release of chlorine atoms, scientists now also realize that bromine-containing halons used in fire-extinguishing materials deplete ozone even more efficiently and are responsible for about 20% of total ozone destruction (Logan, 1990). Other substances like methyl chloroform and carbon tetrachloride (from industrial chemicals and oceans), also contribute to ozone depletion by the release of chlorine, as do large volcanic eruptions and the burning of rainforests.

Research is continuing to search for better substitutes for CFCs. Even if global production of CFCs completely ceased, the atmosphere concentration of chlorine atoms would continue to rise in the next few decades as the gases stored in old appliances gradually leaked into the air. It takes hundreds of years to reduce the concentration of chloride gases in the stratosphere (Logan, 1990).

The success of the Montreal Protocol depends on the participation of all nations. Many developing countries have just begun to use CFC-based technologies and some countries are reluctant to give up the short-term benefits of using CFCs. The replacements for CFCs are generally more expensive and less effective than the original compounds and therefore adjustments to alternatives will not be straight-forward.

1.3 Loss of biological diversity

1.3.1 Context

The term biodiversity is short for biological diversity and describes the 'variety and variability of genes, species, populations and natural ecosystems' (UNCED, 1996, p. 1). Broadly speaking it refers to the number of species that can be found in nature. The exact number of organisms is not known but it is believed to be approximately 5 to 30 million, of which only 1.4 million have been described

(Wilson, 1989). The loss of biodiversity has and continues to occur as a result of habitat destruction, over harvesting, pollution, and the inappropriate introduction of foreign plants and animals.

Most of the species are found in the rainforests, coral reefs, geologically ancient lakes and coastal wetlands. It is estimated that rainforests cover only 7% of the Earth's land surface but they contain almost half the total species count. The UNCED recognizes biological resources as a major capital asset and promotes the development of baseline surveys of resources including identification of the potential economic and social benefit that these resources present.

Biological diversity can be classified into three levels; namely genetic diversity, species diversity and ecosystem diversity (Beder, 1993). Genetic diversity is the sum of genetic information contained in the genes of individual plants, animals and micro-organisms. Species diversity refers to the number of different living organisms on Earth. Ecosystem diversity (also called habitat diversity) relates to the variety of different habitats within an ecosystem.

Biodiversity is important in many ways. First, it sustains food production. With an increasing rate of growth in world population, the demand for food becomes critical. Second, species are the source of medicines to cure a range of known diseases, as well as for medical research and development (Bates, 1990). Third, rainforests play an important role in the planetary recycling of carbon, nitrogen and oxygen by helping to regulate the greenhouse effect by absorbing carbon dioxide out of the air and returning oxygen to the atmosphere. Fourth, the planet is an interwoven ecosystem. The existence of one species is important to the existence of another. The demise of one species may eventually lead to the loss of many others dependent upon it, which may result in accelerated loss of important genetic information (Bates, 1990).

1.3.2 Deforestation

The greatest immediate threat caused by deforestation is the loss or disruption of natural habitats. Deforestation has been an ongoing human activity that has significantly increased since the Industrial Revolution. In the year 900, 40% of the Earth was covered by forest. In 1900, the forested area was reduced to 30%, a reduction of 10% over one thousand years. Today only 20% of the Earth's surface is forested, another 10% reduction for just one hundred years and the remaining forests are decreasing at an accelerating rate (Bates, 1990).

The soils of tropical rainforests are often too infertile to support most forms of human agriculture. They are almost impossible to farm because these regions are not periodically glaciated. The ashes and decomposing vegetation release a flush of nutrients adequate to support the growth of agricultural crops for two to three years and after that artificial supplements are necessary to carry out intensive farming (Wilson, 1989).

Deforestation has begun to disrupt the Earth's equilibrium. The loss of such a large portion of the rainforest means that less sunlight will be absorbed and more stored carbon will be released into the air. This is one of the major causes of global warming. Since 1860, forest clearing has released about 100 to 200 million tonnes of carbon into the atmosphere (Wilson, 1989). This affects the global climate as more carbon dioxide released into the atmosphere raises overall temperatures.

The major barrier that presents itself in combating deforestation lies in the complex multiple role that forests play to society and environment. Socially, forests provide an important 'free' economic resource for many nations while simultaneously providing shelter and protection. In developed nations they also present an important cultural and recreational dimension to society. Unfortunately, the exploitation of forest resources has lead to a tripartite dilemma. Examples of this dilemma include the indigenous Amazon Indians and their plight to save their virgin forests while at the same time being dependent upon the sale of timber for their economic livelihood. The forest burning in Indonesia as a cheap and fast method of land clearing and most recently the beech harvesting on the West Coast of New Zealand are also examples.

The need for timber and industry employment must be balanced against conservation through effective resource management. Issues that arise include the conservation of forests in representative ecological systems and landscapes, the maintenance of primary old-growth forests, the conservation and management of wildlife, the conservation of genetic resources, and the conservation of biological diversity and traditional habitats of indigenous people, forest dwellers and local communities (UNCED, 1996).

Loss of biological diversity and species extinction are obviously linked. Recent analyses predict the loss of species due to deforestation will be about 2 to 8% of all existing species over the next 25 years and the extinction rates are accelerating (Wilson, 1989). The genetic erosion is now recognized as a major threat to the long-term maintenance of global food supplies. The rapid loss of biological diversity has become the subject of increasing national and international concern.

1.4 Population growth

1.4.1 Rates of growth

World population has more than doubled from 2.52 billion in 1950 to 5.29 billion in 1990 and is estimated to exceed 6 billion early this century (UNEP, 1999). It is estimated that 90% of the increase in population occurs in developing countries (Botkin and Keller, 1995) and is a result of the birth rate rather than greater life expectancy or advancements in medicine. Population increases at an exponential rate and places a corresponding demand on food production. Nevertheless, the rate of population growth has fallen from that predicted as little as 20 years ago largely due to progress in reducing fertility rates globally.

'Currently, the highest fertility rates tend to be found in countries suffering from poverty, food insecurity and natural resource degradation' (UNEP, 1999, p. 6). For the most part Asia continues to provide the greatest challenge to global sustainability. Over 50% of the world's population is in Asia and it also has the highest predicted population growth rate of 1.9% or twice that of the global average.

A strong correlation exists between high fertility rates and poverty which both exacerbate environmental stress. The main principle behind the United Nations Conference for Environment and Development *Agenda 21* is that eliminating poverty will result in reduced environmental degradation. The problem is that growth in wealth leads to increased consumption and then to creation of different environmental problems. Thus the promotion and implementation of sustainable measures in the expansion of developing nations is seen as paramount to future global health.

The depletion of soil fertility and water reserves is due to population growth and associated increased crop production and over-farming. In order to maintain productivity of the soil, farmers have to use chemical fertilizers. Traditionally farmers rotated crops to build up soil and control pests. However, due to increased population density and demand for food supply, farmers have shortened the rotational cycle and prevented forest regeneration. Crops that are dependent on chemical fertilizers tend to rob the soil of its natural fertility, which in turn will require more fertilizers in succeeding years. After a number of years overall productivity declines and so even more fertilizers are applied to compensate.

The need for greater fertilization for farming has further sped up the rate of global warming. Fertilizer production involves mining and processing of phosphate and nitrogen-bearing ores, and this process consumes fossil fuels and releases carbon dioxide, methane and other greenhouse gases. Fertilizers also reduce the ability of soil microbes to remove carbon from the atmosphere.

An example of this is seen in the world grain harvest. In 1987 it was 459 million tonnes which was enough to feed the world for 101 days. However, in 1989, the world grain harvest had reduced to only feed the world for 60 days, a reduction of about 41% in two years (Bates, 1990). The reason for the decline in production is due in part to the depletion of the environment. As population grows, greater demands are placed on land use, leading to deforestation, loss of biodiversity, water resource shortage, waste of natural resources, loss of soil fertility and increased soil erosion. This is especially serious in the developing countries where deforestation is at its highest.

The growth of human population is the most significant threat to the global environment. The demands that more people place on natural resources will lead to lower standards of living, more pollution, climate change and loss of biological diversity. Ultimately there can be no long-term solutions to environmental problems unless there is a cessation in the growth of human population. Even the most optimistic forecasts indicate a doubling of population during this century.

1.4.2 Pollution

Pollution is essentially an anthropogenic process and can be characterized as water pollution, air pollution and waste disposal on land. Examples of pollution include sewage, smog, hazardous waste, fertilizers and garbage. Pollution becomes a problem only when the natural assimilative ability of ecosystems is reached. In major population centres evidence of this problem is clear. Technology and development contribute to pollution, supplemented by other events such as major oil spills and nuclear accidents, and threaten the ability to deliver improvements in living standards in the future. Fresh water supplies and clean air to breathe are obviously critical to long-term human health.

There are few solutions to pollution other than prevention through better production methods and a greater realization of the problem at all levels of society. Yet in the context of increasing world population any significant improvement in environmental quality is unlikely.

1.5 Conclusion

Clearly environmental problems are interrelated, and the planet needs to be seen as a whole rather than a series of separate parts that need attention. It is fair to say that the planet is in environmental crisis, in the context that future generations may inherit less than their right because of past and present mismanagement. The environmental crisis is also an economic crisis. It is caused by economic activities and it undermines the very functions on which the economy depends.

References and bibliography

Bates, A.K. (1990). *Climate in Crisis*. The Book Publishing Company.

Beder, S. (1993). *The Nature of Sustainable Development*. Scribe Publications.

Bon, R. and Pietroforte, R. (1999). 'The Italian Residential Construction Sector: An Input-Output Analysis'. *Construction Management and Economics*, vol. 17, pp. 297–303.

Botkin, D. and Keller, E. (1995). *Environmental Science: Earth as a Living Planet*. John Wiley & Sons.

Bowers, J.K. (1997). *Sustainability and Environmental Economics: An Alternative Text*. Longman.

Clayton, A.M.H. and Radcliffe, N.J. (1996). *Sustainability: A Systems Approach*. Earthscan Publications.

Dovers, S. (1994). *Sustainable Energy Systems: Pathways for Australian Energy Reform*. Cambridge University Press.

Dragon, A.K. and Jakobsson, K.M. (eds.) (1997). *Sustainability and Global Environmental Policy: New Perspectives*. Edward Elgar Publishing.

Falk, J. and Brownlow, A. (1989). *The Greenhouse Challenge: What's to be Done*. Penguin Books.

Field, B.C. (1997). *Environmental Economics: An Introduction* (2nd edition). McGraw-Hill.

Firor, J. (1990). *The Changing Atmosphere: A Global Challenge*. Yale University Press.

Fossdal, S. (1999). *Review of Atmospheric Emissions for GBC International Framework Committee*. Norwegian Building Research Institute.

Gilpin, A. (2000). *Environmental Economics: A Critical Overview*. John Wiley & Sons.

Gribbin, J. (1988). *The Hole in the Sky: Man's Threat to the Ozone Layer*. Corgi Books.

Gupta, A. and Asher, M.G. (1998). *Environment and the Developing World: Principles, Policies and Management*. John Wiley & Sons.

International Building Research Council (1999). 'Agenda 21 on Sustainable Construction', *CIB Report Publication 237*, The Netherlands.

IPCC (1995). Information available at <http://www.usgcrp.gov:80/ipcc/>

Kenny, M. and Meadowcroft, J. (eds.) (1999). *Planning Sustainability*. Routledge.

Kirkby, J., O'Keefe, P. and Timberlake, L. (1995). *The Earthscan Reader in Sustainable Development*. Earthscan Publications.

Kolstad, C.D. (2000). *Environmental Economics*. Oxford University Press.

Leary, N.A. and Scheraga, J.D. (1994). 'Policies for the Efficient Reduction of Carbon Dioxide Emissions', *International Journal of Global Energy Issues*, vol. 6, no. 1-2, pp. 102–111.

Levine, M.D., Price, L. and Martin, N. (1995). *Efficient Use of Energy Utilizing High Technology: An Assessment of Energy Use in Industry and Buildings – Summary of Findings*. World Energy Council, Industrial Energy Research.

Logan, R.A. (1990). *Environmental Issues for the 90s*. Environmental Reporting Forum.

Lovelock, J.E. (1979). *Gaia: A New Look at Life on Earth*. Oxford University Press.

Nagle, G. and Spencer, K. (1997). *Sustainable Development*. Hodder & Stoughton.

Nath, B., Hens, L. and Devuyst, D. (1993). *Instruments for Implementation: Environmental Management* (volume 3). Vubpress.

New South Wales Government (1990). *A Greenhouse Strategy for New South Wales*. New South Wales Cabinet Committee on Climate Change, Sydney.

Opschoor, J.B., Button, K.J. and Nijkamp, P. (eds.) (1999). *Environmental Economics and Development*. Edward Elgar Publishing.

Russell, P. (1998). 'Energy Related Environmental Impacts of Buildings – IEA Annex 31', preamble sourced from author.

Santos, M.A. (1999). *The Environmental Crisis*. Greenwood Press.

Schipper, L. and Meyers, S. (1994). *Energy Efficiency and Human Activity: Past Trends and Future Prospects*. Cambridge University Press.

Tietenberg, T. (1998). *Environmental Economics and Policy* (2nd edition). Addison-Wesley.

UNCED (1996). *Habitat Agenda 21*. United Nations Conference for Environment and Development <http://www.infohabitat.org/agenda21>

Underdal, A. (ed.) (1997). *The Politics of International Environmental Management*. Kluwer Academic Publishers.

UNEP (1999). *Global Environment Outlook 2000*. United Nations Environment Programme, Earthscan Publications.

Wilson, E.O. (1989). *Biodiversity*. National Academy Press, Washington DC.

Sustainable development

2.1 Introduction

The risks of treating economic management and environmental quality as if they are separate, non-interacting elements have now become apparent. The world cannot continue to deplete the stratospheric ozone layer by indiscriminate use of chlorofluorocarbons (CFCs). Furthermore, damage to the ozone layer affects human health and economic productivity. Few would argue now that we can perpetually postpone taking action to contain the emission of greenhouse gases (GHGs). Our use of fossil fuels is driven by the goals of economic change and that process will affect global climate. In turn, global warming and sea-level rise will affect the performance of economies.

Sustainable development is not a new idea. Many cultures over the course of human history have recognized the need for harmony between the environment, society and economy. What is new is an articulation of these ideas in the context of a global industrial and information society.

Sustainable development focuses on improving the quality of life for all of the Earth's citizens without increasing the use of natural resources beyond the capacity of the environment to supply them indefinitely. It requires an understanding that inaction has consequences and that innovative ways must be found to change institutional structures and influence individual behaviour. It is about taking action, changing policy and practice at all levels from the individual to the international.

In 1962 *Silent Spring* by R. Carson was published, a book many consider a turning point in our understanding of the interconnections between the environment, economy and social well-being. In the decades that have followed, many milestones have marked the journey toward sustainable development.

2.2 Historical context

The concept of sustainable development gained momentum in the 1980s when scientific evidence about depletion of the environment became obvious. It is now widely recognized that environmental quality and the conservation of natural resources are of importance for the well-being of humankind today and for generations to come.

This recognition was first discussed in the 1973 United Nations Conference on the Human Environment in Stockholm. Furthermore, the ideas of sustainable development have been discussed in the 1980 World Conservation Strategy (WCS), produced by the International Union for Conservation of Nature and Natural Resources (IUCN) in collaboration with the United Nations Environmental Programme (UNEP) and the World Wildlife Fund (WWF, now the World Wide Fund for Nature). Based on the World Conservation Strategy the National Conservation Strategies for sustainable development were then prepared and adopted by the governments of fifty countries. However, the World Conservation Strategy has had little practical impact.

The World Commission on Environment and Development (WCED), created by the United Nations as a result of a General Assembly Resolution in autumn 1983, published a report on sustainable development called Our Common Future in 1987. The report incorporated the concept of sustainable development presented by the World Conservation Strategy in 1980. The Commission was chaired by Mrs Gro Harlem Brundtland, Prime Minister of Norway, and the report became known as the Brundtland Report and has since been widely accepted and quoted.

By the end of 1988 Our Common Future had received public backing from the leaders of the world. The Earth Summit was held in Rio de Janeiro in 1992 by the United Nations Conference on Environment and Development (UNCED) to follow up on key recommendations and strategies previously identified in the Brundtland Report. This conference, also known as the Rio Conference, was the first time the discussion of the planet's future was attended by world leaders. The Earth Summit reached a number of important conclusions, and the Rio Declaration on Environment and Development was agreed setting out 27 general principles for achieving sustainable development. To support this declaration the Summit adopted Agenda 21, which is an action plan to pursue the principles of sustainable development into the twenty-first century. The establishment of a new Commission on Sustainable Development (CSD) under the aegis of the United Nations was established to monitor progress.

In December 1997 the Kyoto Climate Summit was held to set targets for greenhouse gas emissions. The conclusions of the Kyoto Summit were that the industrialized countries, known as Annex 1, were to bring their collective greenhouse gas emissions down by at least 5% below 1990 levels by 2008–2012. That is a considerably weaker target than is necessary to avoid the risk of dangerous climate change. The Protocol was meant to put in place legally binding targets and timetables for industrialized countries to reduce their emissions of greenhouse gases. However, many loopholes exist in the new treaty and it could take several more negotiations before the rules and regulations of the issues put forwarded in Kyoto are environmentally sound, agreed by all the countries and ratified (WWF, <http://panda.org>).

2.3 Definition

More than 70 definitions of sustainable development have been made and used or interpreted by different groups to suit their own goals. Definitions of sustainable development abound since what constitutes development for one person may not be development or progress for another. Development is a value word: it embodies personal ideals, aspirations and concepts of what constitutes good for society. Yet in all the writing on sustainable development there is a common thread, a fairly consistent set of characteristics that appear to define the conditions for sustainable development that need to be achieved. The most popular definition is the one given in the Brundtland Report:

> 'development that meets the needs of the present without comprising the ability of future generations to meet their own needs' (cited in Kirby et al., 1995, p. 1).

This means that for an economic activity to be sustainable it must neither degrade nor deplete natural resources nor have serious impacts on the global environment inherited by future generations. For example, if greenhouse gases build up, ozone is depleted, soil is degraded, natural resources are exhausted and water and air are polluted, the present generation clearly has prejudiced the ability of future generations to support themselves.

Development embodies a set of desirable goals or objectives for society. These goals undoubtedly include the basic aim to secure a rising level of real income per capita and hence increase the standard of living. But it is now accepted there is more to development than economic growth. While economic growth is defined as an increase in real income per capita, development simply means desirable change. Whether growth constitutes development depends on social goals. An increase in real income per capita may be one in a list of development objectives, along with reductions in crime, improved educational and cultural opportunities, greater employment security, increased racial and gender equality, or clean beaches and scenic landscapes (Zarsky, 1990).

The means of achieving sustainable development are identified by Pearce et al. (1989) as consisting of:

1. *Environmental value*. Sustainable development typically involves a substantially increased concentration on the real value of the natural, built and cultural environments. This higher profile arises either because environmental quality is generally seen as an important factor contributing to the success of the more traditional development objectives such as rising real incomes, or simply because environmental quality is increasingly being viewed as part of the wider development objective and as instrumental in the achievement of an improved quality of life.
2. *Futurity*. Sustainable development involves a concern not only for the short to

medium term time horizon, but also for the longer term which will ultimately impact on the inheritance of future generations and their quality of life.

3. *Equity*. Sustainable development places emphasis on providing for the needs of the least advantaged in society (intragenerational equity) and on a fair treatment of future generations (intergenerational equity).

Sustainable development therefore deals with three primary concepts: environment, futurity and equity. These concepts are integrated in sustainable development through a general underlying theme that future generations should be compensated for reductions in the endowments of resources brought about by the actions of present generations.

Munasinghe (1993) suggests economic, ecological and socio-cultural as three approaches to sustainable development. The economic approach to sustainable development is to maximize the flow of income while maintaining the stock of assets (or capital). The ecological approach is to protect biological and physical systems. The socio-cultural concept of sustainable development is to stabilize the social and cultural systems and to minimize destructive conflicts for both intragenerational and intergenerational equity. From a practical point of view the renewable resources such as scarce natural resources should be utilized at rates less than or equal to the natural rate of regeneration. Non-renewable resources should be optimized subject to an ability for substitution between these resources and technological progress. Waste production should be minimized and recycling or reuse maximized.

2.4 The Brundtland Report and sustainable development

The Brundtland Report is considered to be anthropogenic, or human-centred, which means that it is primarily concerned with human welfare through meeting needs and ensuring quality of life over and above protection of the environment (Kirby et al., 1995). The Brundtland Report does not guarantee the needs or quality of life of animals or other living organisms, except in so much as this will benefit humankind. However, to ecologists, living organisms have a right to exist regardless of whether they are beneficial or valuable to humans.

The Brundtland Report suggests that equity can be used to overcome environmental problems. This means that inequality between the developed and developing countries has to be dealt with by raising the living conditions in the developing countries that are generally impoverished.

The Brundtland Report suggests seven strategies to sustainable development:

1. Reviving growth.
2. Changing the quality of growth.
3. Meeting essential needs for jobs, food, energy, water and sanitation.
4. Ensuring a sustainable level of population.
5. Conserving and enhancing the resource base.
6. Reorienting technology and managing risks.
7. Merging the environment and the economy in decision-making.

The Rio Conference was used to follow up the strategies set by the Brundtland Report and mainly concerned conventions on biodiversity, climate change and principles of forest management.

2.5 The environment and economic growth

Economic growth is an ultimate goal for every government to achieve in order to improve standards of living and increase the capacity of providing goods and services to satisfy human needs. Little attention has therefore been paid to the environment. In the 1970s people argued that economic growth put too much pressure on the consumption of renewable and non-renewable natural resources and was directly causing environmental depletion (Beder, 1993).

In the past attention has concentrated on the environment for raw materials and natural resources to carry out economic activities such as agriculture, mining and manufacture, and to provide food, shelter and clothing for human needs. The effects of the environment upon economic systems were largely ignored. Today people recognize that environmental degradation can, in turn, affect economic activities and the ability to meet development goals.

Munasinghe (1993) describes the environment as providing three main types of services to human society. First, natural resources provide raw materials as required by humans for economic activities. Second, the environment acts as a sink to absorb and recycle the waste products of economic activities (such as forests extracting carbon dioxide from the air and returning oxygen to the atmosphere). Third, it is recognized that the environment supports life on Earth. Economic activities, therefore, are greatly affected if the environment is degraded or the resource base is significantly diminished.

Economic growth and environmental protection have a two-way interaction. Economic growth is a necessary part of sustainable development (Beder, 1993). Sustainable development, therefore, is about ensuring that economic activity that supplies communities with food, shelter, manufactured goods and services for today can be continued into the future.

According to Carley and Christie (1992), sustainable growth of society refers to the carrying capacity or the number of people that can be supported at any time on a sustainable basis. In other words, the sustainable growth of an economy is the maximum rate of resource consumption and waste production that can be sustained indefinitely in a region without impairing ecological productivity and integrity.

Beder (1993) suggests that economic growth can be achieved in a number of ways, some of which have more impact on the environment than others. Sustainable development aims to achieve economic growth by increasing productivity through technological change without excessively increasing natural resource use.

2.6 Principles of sustainability

To demonstrate the link between economics and the environment it is first necessary to define sustainability in both economic and ecological terms. Economic and ecological sustainability precepts may be considered as partially overlapping circles, as shown in Figure 2.1. The overlap marks the territory of sustainable development.

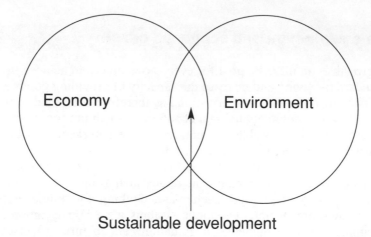

Figure 2.1 Conceptual framework for sustainable development

Zarsky (1990) defines sustainability as the way that humans manage an economy to preserve its productiveness. It can be described by four key attributes:

1. *Efficiency*. Projects implemented and production processes employed should be efficient and therefore should yield the greatest output per unit within the bounds of current technology. For market-based economies, inputs and outputs are measured by their monetary values.
2. *Investment*. The total resource base for production (comprising human, manufactured and natural resources) should not be diminished. Investment should be sufficient to at least replenish and preferably to expand the resource base. While there are short-term consumption gains from depleting the productive stock, in the long term depletion destroys the capability for an economy to function. Of course, investment requires that the economy generates an investible surplus.
3. *Diversification*. Sources of inputs and the range of outputs should be diversified as much as possible so that the system as a whole is less vulnerable to internal or external risks.
4. *External balance*. The value of imports and exports should balance over the long term.

Zarsky (1990) defines ecological sustainability as the way in which humans should interact with the biosphere to maintain its life-support function. It can be described by the following five attributes:

1. *Biodiversity*. All species of flora and fauna and their habitats should be conserved, maintaining the natural potential for species to evolve.
2. *Ecosystem conservation*. The natural stock of ecological resources, such as soil,

ground and surface water, land biomass and water biomass, has regeneration limits. Ecosystems play a vital life-support function and should be protected.
3. *Interconnectedness*. Improvements in environmental quality in one country should not be achieved at the expense of another.
4. *Aversion to risk*. The future is unpredictable and it is best to be cautionary and to make decisions based on avoidance of potentially bad consequences, even if this means that returns are not maximized in the short term. This is particularly important given unknown thresholds wherein incremental change can suddenly give way to sweeping systematic change. Any activity that has the potential to irreversibly change an ecosystem from one state to another should be avoided.
5. *Scale of impact*. Humans should minimize their use of mass and energy flows relative to the total mass and energy flows of the relevant ecosystem.

2.7 Conserving natural capital

The fundamental principle of sustainable development is intergenerational equity. Fulfilling development goals today should not be allowed to jeopardize the ability of future generations to meet their own needs.

There is debate, however, over what exactly should be protected. Some economists argue that natural and human-made capital stocks are interchangeable: one can switch between them and thus trade off some depletion in the natural environment for the enhanced income benefits of increases in plant, equipment or other human-made capital assets. Sustainable development in this context would mean preserving the total asset stock.

Others, such as Pearce et al. (1989), argue that the relevant concept for sustainable development is the non-depletion of the natural capital stock. Future generations should inherit at least a similar natural resource base as present generations were afforded.

There is no doubt that investment in technology and the development of knowledge today will benefit generations in the future. Nonetheless, Pearce (cited in Zarsky, 1990) argues that there are four reasons why natural and human stocks of capital should not be treated together:

1. *Non-substitutability*. Technology can substitute for services provided by the natural environment on a local scale and to a limited extent. Despite this, natural and human-made capital are generally not interchangeable, particularly in terms of global ecosystem functions.
2. *Uncertainty*. Future technological breakthroughs might increase the interchangeability of natural and human-made capital. Ecologists argue that when faced by uncertainty decisions should be made that avoid risk. In other words, natural capital should not be depleted on the assumption that sometime in the future a technological replacement will be created.
3. *Irreversibility*. Some forms of natural environment degradation can never be restored by human technology or ingenuity.
4. *Equity*. Typically, environmental degradation is suffered to a greater extent by the poor in society than those that are wealthy. This situation is no more obvious

than in developing countries where the poor depend heavily on the productivity of natural resources for their survival. Treating natural and human-made capital as interchangeable will disadvantage disproportionately the poorer members of society.

The above reasons suggest that sustainable development should aim to ensure that present economic activity does not deplete the stock of natural capital. Conserving the environment should be treated as a constraint on all economic activity. Within this constraint projects should maximize net economic benefits.

2.8 Sustainable construction

Sustainable construction is perhaps best described as a subset of sustainable development. It focuses on the issues of procurement, assembly and commissioning and embraces matters such as tendering, site planning and organization, material selection, recycling and, in particular, waste minimization.

Procurement processes provide an opportunity for clients to introduce objectives that contractors must meet during site construction. These may comprise environmental performance standards, verification measures and reporting requirements. For example, when a client opts for a 'design and build' process, it is important to specify the level of environmental performance expected, not just for the completed design, but also the activities necessary to deliver that design. In such situations contractors may be required to submit an environmental performance statement as part of their bid, or undertake a full life-cost report, or comply with some other contractual obligation aimed at ensuring the work meets expected standards.

Assembly activities are usually left to the contractor to determine. This is an opportunity for contractors to contribute to environmental performance through site processes that minimize damage and inconvenience to the surrounding landscape. Issues may include noise, site run-off, excavation methods, disposal of building waste, and minimization of material requiring disposal, especially to landfill. Strategies commonly used on large modern sites include waste sorting, recycling and stormwater containment.

Commissioning should entail information provision that enables building users to understand the way the design is intended to operate and how to fine-tune systems to meet user needs into the future. Information should relate to warranties and performance specifications, maintenance cycles and methods, component adjustment, upgrade and system monitoring, and becomes a key quality control tool for a facility manager or building supervisor.

The success of any plan to improve the environmental performance of the site construction process relies on proper training of site operatives and effective supervision. It is recommended that a waste management plan is formally developed to deal with issues such as choice of materials to be recycled, on-site sorting strategies at the point of waste creation, stormwater run-off prevention and the cost implications of decisions. Although the construction process is relatively short-lived in comparison to a building's intended life, it is nevertheless a potentially damaging exercise and one that offers considerable opportunity for improvement.

A large proportion of material sent to landfill is construction related. Sustainable construction requires that building waste, whether comprising new materials or demolished components, is recycled whenever possible and not merely discarded as landfill. Therefore effective waste management and minimization of site waste are the key issues in ensuring that modern construction does not have inappropriate impacts on the environment.

2.9 Conclusion

Sustainable development is a fluid concept that will continue to evolve over time, but there remains a set of common characteristics underlying the many streams of thought. Sustainable development emphasizes the need for understanding the interconnections between the environment, economy and society, taking a long-term view of present actions, and having concern for equity and fairness.

In addition, sustainable development strategies usually highlight the interplay between the local and global, the developing and the developed, and the need for co-operation within and between sectors. Sustainable development is not a detailed plan of action, a formula that can be followed blindly. There is no one solution. Solutions will differ between places and times and depend on the mix of values and resources found there. Approaching decision-making processes from a sustainable development perspective requires undertaking a careful assessment of the strengths of households, communities, companies and governments to determine priority actions.

References and bibliography

Beder, S. (1993). *The Nature of Sustainable Development*. Scribe Publications.

Botkin, D. and Keller, E. (1995). *Environmental Science: Earth as a Living Planet*. John Wiley & Sons.

Bowers, J.K. (1997). *Sustainability and Environmental Economics: An Alternative Text*. Longman.

Carley, M. and Christie, I. (1992). *Managing Sustainable Development*. Earthscan Publications.

Carson, R. (1962). *Silent Spring*. Macmillan.

Clayton, A.M.H. and Radcliffe, N.J. (1996). *Sustainability: A Systems Approach*. Earthscan Publications.

Common, M. (1995). *Sustainability and Policy: Limits to Economics*. Cambridge University Press.

Dragon, A.K. and Jakobsson, K.M. (eds.) (1997). *Sustainability and Global Environmental Policy: New Perspectives*. Edward Elgar Publishing.

Field, B.C. (1997). *Environmental Economics: An Introduction* (2nd edition). McGraw-Hill.

Gilpin, A. (2000). *Environmental Economics: A Critical Overview*. John Wiley & Sons.

Gupta, A. and Asher, M.G. (1998). *Environment and the Developing World: Principles, Policies and Management*. John Wiley & Sons.

Hutchinson, A. and Hutchinson, F. (1997). *Environmental Business Management: Sustainable Development in the New Millennium*. McGraw-Hill.

Kaya, Y. and Yokobori, K. (eds.) (1997). *Environment, Energy and Economy: Strategies for Sustainable Development*. United Nations University Press.

Kenny, M. and Meadowcroft, J. (eds.) (1999). *Planning Sustainability*. Routledge.

Kirby, J., O'Keefe, P. and Timberlake, L. (1995). *The Earthscan Reader in Sustainable Development*. Earthscan Publications.

Kolstad, C.D. (2000). *Environmental Economics*. Oxford University Press.

Munasinghe, M. (1993). *Environmental Economics and Sustainable Development* (Paper Number 3). The World Bank.

Nagle, G. and Spencer, K. (1997). *Sustainable Development*. Hodder & Stoughton.

Opschoor, J.B., Button, K.J. and Nijkamp, P. (eds.) (1999). *Environmental Economics and Development*. Edward Elgar Publishing.

Pearce, D., Markandya, A. and Barbier, E. (1989). *Blueprint for a Green Economy*. Earthscan Publications.

Pearce, D., Barbier, E. and Markandya, A. (1990). *Sustainable Development Economics and Environment in the Third World*. Edward Elgar Publishing.

Santos, M.A. (1999). *The Environmental Crisis*. Greenwood Press.

The Secretaries of State for the Environment (1994). *Sustainable Development: The UK Strategy*. HMSO.

Underdal, A. (ed.) (1997). *The Politics of International Environmental Management*. Kluwer Academic Publishers.

World Commission on Environment and Development (1990). *Our Common Future* (Australian edition). Oxford University Press.

Zarsky, L. (1990). *Sustainable Development: Challenges for Australia*. The Commission for the Future, Canberra.

3

Accounting for the environment

3.1 Introduction

One of the central activities in both environmental economics and sustainable development is the placement of proper values on the services provided by the natural environment. Many of these services currently are perceived as having a zero price simply because no marketplace exists in which their true values can be revealed through the commercial actions of buying and selling. Supply and demand theory indicates that if something is provided at a zero price then more of it will be demanded than if it had a positive price. The danger is that a greater level of demand for the environment will be more than natural systems can sustainably support.

It thus is critical that resources and environments which serve economic functions have a positive economic value. To treat them as if they had a zero value is to seriously risk exploitation and depletion. An 'economic function' is defined as any service that contributes to human well-being, to the standard of living or to development. This simple logic underlies the importance of valuing the environment correctly and integrating those values into a convention (or 'neo-classical') economical framework, and it is in relation to this last objective that discounting assumes great responsibility.

Environmental concerns tend to be dissipated by current generations because their impact will occur well into the future and hence rectification can be undertaken at a later time. One reason for postponing action is that future costs are considered less of a burden than current costs. This is reflected in the discounting process, which plays a fundamental role in economic evaluations. Discounting suggests that deferring problems is preferable and this translates to an emphasis on reactive rather than anticipatory policy, which Pearce et al. (1989) refer to as the 'fix it later' syndrome.

But other considerations work against this purely economic view. Environmental damage can become irreparable (or irreversible) and hence its deference into the future where it will presumably be fixed later might be impossible, even if the relative value of the future money used to undertake the repair might be quite small. Furthermore, environmental issues are often treated as if they represent some minor problem in the otherwise efficient working of an economic system that can easily be overcome through theoretical compensation allowances or else ignored completely. Yet the essential feature about environments is that their workings are pervasive and therefore the effects of damage, depletion, pollution, energy usage, waste disposal and the like may manifest themselves in a variety of unexpected ways. Deferring environmental issues to the future therefore attracts a greater level of uncertainty about the nature of the problem, its effect and its ultimate solution.

3.2 The concept of sustainable income

As population is growing and economic activities are becoming intensified, greater pressure is being put on the existing stock of natural resources. As a result more damage is being caused to the environment and the stock of natural resources is being progressively exhausted. Today the fundamental problem facing the economy is the depletion of renewable resources and the exhaustion of non-renewable resources. More attention is now being paid to the conservation of such resources in order that future generations will be able to enjoy the same natural wealth as that available to the present generation.

Environmental accounting is becoming more important as economists recognize that conventional national accounting cannot reflect the sustainable income of a nation. Capital wealth increases do not truly represent the economic growth and economic success of a country because environmental depletion has been excluded. Economists now argue that a modification to conventional national accounting is necessary in terms of subtracting the cost of damages and changes to the environment from the final figure. Environmental accounting is critical for both investment decisions involving environmental impacts and for the proper regulation of the environment.

Sustainable income is the true income of a nation. According to Daly (1989) sustainable income is the maximum amount that a nation can consume without reducing the amount of possible consumption in a future period. This concept is not just looking at current earnings but also changes in capital. In order to arrive at the principle of sustainable income a number of factors need to be subtracted from the gross national product (GNP). Traditionally, only depreciation for plant and equipment is subtracted to arrive at net national product (NNP) but no consideration is being taken for the depreciation of natural assets.

Daly suggests two adjustments to NNP to move it closer to the concept of sustainable income. First, depreciation for the consumption of environmental assets needs to be included. Second, adjustment also needs to be made for the defensive expenditures such as the cost of environmental protection and the expenditures for environmental damage compensation arising through economic activities. The net national product with the deduction of both defensive expenditure and depreciation of natural capital is the sustainable income of a society.

The measurement of sustainable income allows governments to know the maximum amount of income that can be consumed within a particular period of time without causing its eventual impoverishment. It is therefore important to measure the national income correctly to indicate the true position of a nation.

3.3 Problems with conventional national accounting

Conventional national accounting is a tool used by economists to measure economic activities for a country within a particular period of time. It is a valuable method to indicate short to medium term changes in the level of economic activities of a country. An increase in the national account is often regarded as economic growth and economic success. However, it is less useful for measuring long-term sustainable economic growth as environmental protection costs and the depletion of natural resources is excluded. The outcome is that the economic indicators show the world is prospering but on the other hand the environment is under severe stress as pollution of the environment and exhaustion of natural resources continue simultaneously. It therefore cannot truly reflect sustainable income of a society. The main problems of conventional national accounts are as follows:

1. National accounts only measure goods and services legally bought or sold in the market but the consumption of renewable and non-renewable natural resources are not recorded in the national accounts as they are not sold as commodities. Manufactured assets such as plant and equipment are valued as productive assets and are written off against the value of production as they depreciate. Natural resource assets are not so valued or adequately accounted for. Their loss produces no change in the national account to reflect the decrease in potential future production. Natural resources like water, soil, air and mineral resources are not just raw materials needed to uphold economic activities but also contain waste absorbing and life-supporting capabilities.
2. National accounts may be able to indicate the level of economic activities at a particular period of time but may not accurately represent welfare of the people as environmental degradation has been neglected. People may find that they are living in an area with air and water seriously polluted but the national income indicates that the living standard has been improved.
3. The cost spent on cleaning up the environment is added to the national account and the inflated economy is labelled as exhibiting increased growth and production. It is not adequate that national accounts only incorporate expenditures incurred to compensate the negative consequences caused by economic activities with no record of the consumption of the stock of natural resources.

Due to the problems associated with conventional national accounting, economists argue that national accounts do not reflect social well-being and the true sustainable income of a nation. Therefore, action needs to be taken to incorporate the value of natural assets and expenditure on damage to the environment resulting from economic activities into the national accounts.

Serafy (1989) recommends two approaches to deal with the depletion or degradation of natural resources: the depreciation approach and the user cost approach.

By using the depreciation approach, depletion or degradation of natural resources is needed to be priced or valued and adjustment can be made to GNP to arrive at a correct NNP. The user cost approach avoids the difficulties of putting values on the stock of natural resources. Instead it measures the current extraction rates in relation to the total available stock. No matter what approach is used, it is fundamental to understand and accept the shortcomings of conventional national accounts for indicating true sustainable income.

3.4 Environmental accounting

Environmental destruction is often related to the special properties of environmental goods themselves. To economists, environmental goods are public goods and generate externalities whenever they are utilized. According to Thampapillai (1991), public goods display two properties: non-exclusion and zero marginal costs.

The availability of public goods is often abundant and once they are consumed they do not reduce the quantity available to others. Furthermore, once they are provided they bear no extra cost to additional consumers. To a certain extent the public properties of environmental assets will cease and they will become private goods. Beyond this point people may need to pay for the consumption of environmental goods. Environmental goods are free gifts of nature and their zero price condition has led to them being used excessively, resulting in depletion, deterioration and no incentive for their protection.

As society fails to protect the environment and destruction occurs, these goods become externalities in the market (Beder, 1993). They directly affect the normal functioning of the market and the welfare of society. For instance, manufacturing industries often discharge chemicals into water systems. This results in the deterioration of water quality, poisoning of marine species, destruction of fishing industries and damage to human health. Since this loss cannot be easily measured and valued, it has been ignored in the national accounts. However, even though environmental goods have no market price value, they affect the well-being of humankind both now and in the future.

According to Shechter and Freeman (1994), environmental goods can be characterized into use values (UV) and non-use values (NUV). Use values can be broken down further into direct and indirect use values and option values. Direct use values represent the output of environmental goods that can be consumed directly (such as food, recreation and health). Indirect use values refer to benefits derived from functional services (such as flood control and storm protection). Option values refer to the willingness to pay by consumers to avoid something running out in the future (such as biological diversity and conserved habitats). Non-use values, also known as bequest values, derive from the welfare that people enjoy today and are passed on to future generations.

Both use and non-use values are necessary to be measured and recorded in the national accounts so that people are fully aware of the environmental impacts to the economy as a whole. Environmental goods can be measured by different valuation techniques even though they would not be truly bought or sold in the market.

3.5 Valuation techniques

Economists have developed different techniques to put monetary values on environmental goods. Different valuation techniques have different strengths and weaknesses and the decision on the type of techniques to be used will depend on the nature of environmental goods. The following are some general considerations for making decisions on the type of techniques to be used (OECD, 1994):

1. It is often possible to use more than one type of valuation technique for a proposed project or change. As a result multiple estimates can be established for comparison and greater confidence obtained.
2. Different valuation techniques are used for different types of environmental goods. In this sense they should be considered as complimenting rather than competing with each other. These techniques are not substitutes for one another but rather value different aspects of a proposed project or change.
3. The selection of valuation method depends on individual preferences. In some cases people may have preferences for the use of one technique over another. People do not like survey-based methods because they are considered to be too subjective and unreliable. Therefore, they tend to use methods based on observed market behaviour.
4. The needs of the general public should be considered. When choosing a valuation technique consideration should be given to how the information obtained will be received by the general public.

Economists have developed different techniques to put monetary values on environmental goods. The following are some of the popular methods used on different types of environmental goods.

3.5.1 Willingness to pay method

Willingness to pay (WTP) is also known as contingent valuation (CV). This is a survey-based method used to value environmental goods in the hypothetical market. This hypothetical market must be as close to the real market as possible. The survey is taken in the form of personal interview, a mail survey or a telephone interview. Therefore, the format of the questionnaire is crucial in obtaining accurate results. As recommended by Hoevenagel (1994), the CV questionnaire should consist of three elements: description of the environmental change, method of payment and the description of the hypothetical market.

The method is based on the concept that people have true preferences for all kinds of environmental goods and they are able to transform these preferences into monetary terms. The idea of WTP is to find out how much people are willing to pay to preserve or improve an environment and is based on the assumption that the environment does not already belong to them but they have to buy it (Hoevenagel, 1994).

Another measure of economic value is to ask how much people are willing to accept (WTA) in the way of compensation for environmental damage. Answers obtained through WTP and WTA could differ. In most cases questions about WTA compensation yields higher answers than questions about WTP to retain the same amenity. This tends to be caused by human nature rather than the type of questions being asked. Pearman (1994) suggests that the gap between WTP and WTA can be narrowed with successive iterations.

The advantage of the WTP method is that it can be used to value a variety of environmental goods such as the protection of endangered species, commercial and recreational fishing and health risks (Hoevenagel, 1994; OECD, 1994). This method is the only method available to measure non-use benefits especially for environmental goods that are unique or have few substitutes. However, there are limitations of the WTP method. As Markandya and Richardson (1994) describe, the WTP method has different biases including hypothetical context, information, strategic and payment bias.

Hypothetical context bias arises because it is difficult for people to come out with an accurate figure in a hypothetical market. The figures obtained through WTP or WTA are not actually paid or received. People may tend to inflate or deflate the amount they are willing to pay or accept should they be given the opportunity to buy or sell in the future. Information bias refers to the level of information available to people at the time of the interview and their knowledge about the environment. Strategic bias arises because people tend to understate their willingness to pay as they would prefer to enjoy the benefit at a lower price should they really be given the opportunity to buy. This is often related to a person's income level and their ability to borrow. Finally, payment bias arises because the respondents are sensitive to the method of payment for the improvement that is proposed, such as surcharges, utility fees and taxes.

3.5.2 Stated preference method

This method is also known as contingent ranking. Under this method people are not asked to transform their preferences on environmental goods into monetary values. Instead they are asked to place their preferences on a set of choices of one or more specified multi-attributed alternatives. Then their subjective preferences are inferred and transformed into monetary values.

This technique is used widely for transport planning, air quality, commercial and recreational fishing and valuing the environmental impacts of transport policy (OECD, 1994). Sometimes difficulties may arise due to misunderstanding of the choice context and abandoning one alternative for another.

3.5.3 Does-response method

This method is based on developing a does-response relationship between economic activities and environmental qualities. For many economic activities, environmental quality is considered to be a factor of production. For example, water quality can affect agricultural productivity and industries using water for processing purposes. Therefore, water quality or the cost to improve water quality can be considered as a factor of production in these economic activities.

This method is useful to alert people about the effects of pollution upon their health and economic activities and is a useful tool for decision-makers on the formation of environmental policies. However, this method requires collection of sophisticated data that is difficult to obtain. It can also underestimate the benefits of environmental policy if the indirect use and non-use values are expected to be crucial (Hoevenagel, 1994).

3.5.4 Opportunity cost method

Opportunity cost is often related to the conflicts between preservation and development (Beder, 1993). This method is used to put values on a list of alternative activities that could take place in a preserved area. Benefits will then be worked out for individual activities. Net benefits will be obtained by subtracting the cost incurred in getting these benefits. The highest amount of net benefits will be taken as the opportunity costs of preserving an environment. The opportunity cost method can help the decision-maker and the general public to be aware of the worth of an environmental loss before any decision is made.

3.5.5 Hedonic price method

The hedonic price method is based on the assumption that environmental assets can be valued by estimating the prices of their closest market substitutes (Beder, 1993). For instance, property value will be higher if it contains special features such as a swimming pool, tennis court and access to public transport. However, other environmental qualities can also affect the property value, such as clean air and views. Given the same type of property people may wish to pay more for one with better environmental quality. The differences between the properties will reflect the value the market puts on clean air and views. This method attempts to identify the effect of environmental quality upon value and infer how much people are willing to pay for an improvement in environmental quality.

This method is commonly used for measuring the benefits of changes in environmental risks to human lives, urban air quality, water quality and wildlife hunting (Hoevenagel, 1994). Economists agree that since the hedonic price method obtains results by observing market behaviour, it is more accurate than survey-based methods. However, this method is based on the assumption that people are fully aware of the effects of environmental quality, they can find the exact market substitutes and their WTP prices are the highest. This, in reality, never happens. The method tends to obtain poor results if the seriousness of environmental quality cannot be directly perceived or is unclear to people. Also the method cannot be used to measure non-use values.

3.5.6 Travel cost method

This method uses information on the amount of time and money people incur to access a recreational site. The value of that site can be obtained by estimating their willingness to pay for the facilities available at the site. It is used to value benefits of improvements in recreational facilities in parks, or the values of cultural sites that are visited by people from many different locations. It is used widely by developing countries to estimate benefits from tourism development. This method is costly to implement and only appropriate to valuing recreational benefits where the sites concerned are visited specifically for recreational purposes.

It uses journey costs as substitute values for measuring willingness to pay for the recreational or educational experience. The disadvantage of the travel cost method is people may not take the most direct route to the location. Also recreational activities are often complex and may have more than one activity in the same trip (such as shopping and visiting friends). The results obtained may, therefore, be distorted.

3.5.7 Averting behaviour method

This method is based on the assumption that people are aware of the effects of environmental quality to their well-being. The environmental quality value is estimated by obtaining the costs of expenditure people make to avoid the effects of environmental hazards. For example, due to the depletion of the ozone layer, people have to spend more on sunscreens, hats and protective clothing and the expenditures on health and medical care are higher if the environment is polluted (Hoevenagel, 1994).

This method obtains results through observed market behaviour and is regarded as more accurate than results from surveys. However, it is important to obtain data with regards to the relationship between health, averting behaviour undertaken and its price. It is often difficult to obtain accurate information in this regard. Also this method cannot be used for measuring non-use values.

3.6 Conclusion

Many people argue that putting a price on environmental quality is not useful for protection as none of the valuation methods can accurately value the environment. Furthermore, ecologists argue that matters such as biological diversity cannot be priced at all since plants and animals have an intrinsic value that cannot be represented in monetary terms.

No matter what techniques are to be used to value environmental assets, the primary aim is to ensure that renewable resources are being restored and the rate of consumption of non-renewable resources is slowed to enable an orderly transition to new solutions. Even though putting a price on the environment cannot save the environment, to a certain extent it allows the decision-maker and the general public to realize the damage that projects may cause and, in the process, highlight the importance of environmental conservation.

References and bibliography

Ahmad, Y.J., Serafy, E.S. and Lutz, E. (1989). *Environmental Accounting for Sustainable Development*. The World Bank.
Beder, S. (1993). *The Nature of Sustainable Development*. Scribe Publications.
Bowers, J.K. (1997). *Sustainability and Environmental Economics: An Alternative Text*. Longman.
Canadian Institute of Chartered Accountants (1993). *Environmental Costs and Liabilities: Accounting and Financial Reporting Issue*.
Coenen, F.H.J.M., Huitema, D. and O'Toole, L.J. Jr. (eds.) (1998). *Participation and the Quality of Environmental Decision Making*. Kluwer Academic Publishers.
Coker, A. and Richards, C. (1992). *Valuing the Environment: Economic Approaches to Environmental Evaluation*. Belhaven Press.
Common, M. (1995). *Sustainability and Policy: Limits to Economics*. Cambridge University Press.
Daly, H.E. (1989). 'Towards a Measure of Sustainable Social Net National Product'. In *Environmental Accounting for Sustainable Development* (Y.J Ahmad, S.E. Serafy and E. Lutz, eds.), pp. 8–9, The World Bank.
Field, B.C. (1997). *Environmental Economics: An Introduction* (2nd edition). McGraw-Hill.
Gilpin, A. (2000). *Environmental Economics: A Critical Overview*. John Wiley & Sons.
Gray, R., Beddington, J. and Walters, D. (1993). *Accounting for the Environment*. Paul Chapman Publishing Ltd.
Hanley, N., Shogren, J.F. and White, B. (1997). *Environmental Economics in Theory and Practice*. Macmillan.
Hoevenagel, R. (1994). 'A Comparison of Economic Valuation Methods'. In *Valuing the Environment: Methodological and Measurement Issues* (R. Pethig, ed.), pp. 251–70, Kluwer Academic Publishers.
Kolstad, C.D. (2000). *Environmental Economics*. Oxford University Press.
Laird, J. (1991). 'Environmental Accounting: Putting a Value on Natural Resources'. *Our Planet*, vol. 3, no. 1, pp. 16–18.
Lutz, E. (1993). *Towards Improved Accounting for the Environment*. The World Bank.
Markandya, A. and Richardson, J. (1994). *The Earthscan Reader in Environmental Economics*. Earthscan Publications.
McCornell, R.L. and Abel, D.C. (1999). *Environmental Issues: Measuring, Analysing and Evaluating*. Prentice Hall.
OECD (1994). *Project and Policy Appraisal: Integrating Economics and Environment*. Organization for Economic Co-operation and Development.
Opschoor, J.B., Button, K.J. and Nijkamp, P. (eds.) (1999). *Environmental Economics and Development*. Edward Elgar Publishing.
Pearce, D.W., Markandya, A. and Barbier, E.B. (1989). *Blueprint for a Green Economy*. Earthscan Publications.
Pearman, A. (1994). 'The Use of Stated Preference Methods in the Evaluation of Environmental Change'. In *Valuing the Environment: Methodological and Measurement Issues* (R. Pethig, ed.), pp. 229–49, Kluwer Academic Publishers.
Serafy, S.E. (1989). 'The Proper Calculation of Income from Depletable Natural Resources'. In *Environmental Accounting for Sustainable Development* (J.Y. Ahmad, S.E. Serafy and E. Lutz, eds.), pp. 10–18, The World Bank.
Shechter, M. and Freeman, S. (1994). 'Non-use Value: Reflection on the Definition and Measurement'. In *Valuing the Environment: Methodological and Measurement Issues* (R. Pethig, ed.), pp. 171–94, Kluwer Academic Publishers.
Thampapillai, D.J. (1991). *Environmental Economics*. Oxford University Press.
Tietenberg, T. (1998). *Environmental Economics and Policy* (2nd edition). Addison-Wesley.

Triggs, A.B. and Dubourg, W.R. (1993). 'Valuing the Environmental Impacts of Opencast Mining in the UK: The Case of the Trent Valley in North Staffordshire'. *Energy Policy*, November.
Wills, I. (1997). *Economics and the Environment*. Allen & Unwin.

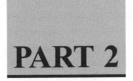

PART 2

Development controls

In an ideal world, sustainable practices would be sought out as far as practicable because conserving resources (present or future) would be the obvious thing to do. However in the real world, where some resources are shared and without monetary labels, development decisions are often made on the basis of profit maximization to the investor within a short-term horizon. This myopic behaviour may dictate choices that are indifferent to sustainable goals, consuming resources that are cost-effective with little concern for long-term implications and wider social objectives.

Even for public sector projects, which may be generally thought of as more community-orientated with emphasis on collective utility and amenity rather than short-term financial gain, monetary evaluation is still dominant. New projects are chosen if their contribution to society is positive and higher than alternative courses of action, and are often assessed using a discounted cash flow approach. Money is typically the adopted metric for development decisions.

Concern over environmental quality has led to the introduction of policies and procedures that help to protect natural systems from the disruptive effects of uncontrolled development. One such mechanism, known as environmental impact assessment, is a process of examining the probable effects of development on local ecosystems and establishing constraints on the developer so that irreversible damage does not occur. Environmental impact statements, therefore, set the parameters within which development is permitted.

Controls exist in some form for all types of development. At the macro level they aim to ensure that society's interests are protected and that concern is exercised for the rate at which environmental goods and services are consumed. At the micro level they aim to ensure that infrastructure is constructed according to recognized standards and accepted levels of occupant safety. Development controls are administered by government authorities but are not applied universally to all new development.

Large-scale projects that are perceived to potentially have a significant effect on their environment are generally required to undergo a formal assessment of impact. However, this approach is not common for smaller projects and virtually non-existent for residential development. Yet collectively these smaller developments have the greatest impact on environmental quality. The avoidance of environmental assessment for some development weakens the effectiveness of controls at a macro level and fails to guarantee the pursuit of sustainability goals.

Incentives are one method of introducing self-regulation for development that is otherwise too small to warrant formal assessment procedures. For example, providing a cash rebate for the cost of home insulation would routinely encourage better insulation performance, reducing heating and cooling demands not only for the occupants but across whole communities. The cost to government of insulation would be offset by savings in electricity generation, which would in turn reduce greenhouse gas emissions and help to protect the environment from the damaging effects of global warming and climate change. Incentives of this type are highly effective for new construction, and can even be extended to progressively upgrade the large stock of buildings already constructed.

The best incentive of all is economics. If environmentally-friendly solutions were also the most cost-effective, a free and properly informed market would seek out such solutions by choice. However, sadly this is not the case. Although there are many examples where it is true, the vast majority of products that exhibit higher performance have a higher price tag attached to them. Currently decisions are needed that involve spending more money now to reap an operational saving over a number of years, but for a variety of reasons decisions are often based on short payback periods and so cheaper products with lower performance remain popular.

But not all aspects of environmental protection are effectively handled by incentives. It is necessary, therefore, that governments increase levels of environmental compliance through the introduction of legislative constraints. Examples may be maximum energy consumption limits per annum for various types of development, including energy involved in mining and manufacture of raw materials, restrictions on waste disposal, contributions to environmental regeneration projects proportional to land usage, and systems of carbon taxes and emission permits.

With legislation comes enforcement and penalties for non-compliance. A regime can easily be established that encourages development to be designed to meet only the set minimum standards, whereas in a free market it might be argued that higher levels of compliance would voluntarily occur. The 'carrot or the stick' debate will continue as long as society as a whole fails to deliver projects which approach sustainable development goals on a routine basis. It is envisaged that environmental legislation will continue to expand, particularly in developed countries, in response to the worsening environmental crisis.

The chapters in this part deal with the necessity to impose limits on environmental resource usage. Chapter 4 discusses the effectiveness of environmental legislation at local, national and international levels. Chapter 5 investigates environmental impact assessment processes and why they form an important weapon against undesirable development. Chapter 6 looks at the wider implications of environmental policy and some of the directions in which it is heading.

Development controls are necessary. As society moves towards limits for a whole range of natural resources, strengthening of controls will surely take place. The degree to which new development is constrained will be a function of our ability to act responsibly and to pursue sustainable development objectives wherever possible. Governments have an obligation to protect society from itself.

Environmental law

4.1 Introduction

The rapid deterioration of the global environment has drawn the attention of people nationwide and worldwide. This awareness of the need for environmental protection and sustainable practices has led to the formulation of statutes and policies as well as guidelines for protecting the environment and identifying solutions to environmental problems.

With increasing global concern over the need to protect the environment, environmental law is becoming an important and rapidly expanding discipline. The development of environmental law is critical to the success of meeting the global challenge to protect the environment. It is not easy to define a boundary for environmental law in the same way as other well-established areas such as contract law. Indeed environmental law is still in a state of development and is undergoing constant refinement in the courts. With a legislative environmental regime comes the need for enforcement and penalties for illegal behaviour, and a considerable social cost to police it.

Bates (1995) describes environmental law as the initiation of administrative structures to protect the environment and the species living in it. It helps to control and manage human activities that have significant impacts on the environment. Through this mechanism people will further realize the importance of the environment and share the responsibility for its conservation.

4.2 The scope of environmental law

4.2.1 Definition

Bates (1995, p. 3) states that 'any rules or regulations which govern conduct which is likely to affect the environment may fall within the definition of environmental law'. The scope of environmental law is very wide and basically can cover any-thing relating to the environment. The term environment is defined to include the natural environment such as land, air, forest and water, as well as the urban, the built and the social environment and even the work environment (Environmental Defender's Office, 1992). Therefore, environmental law concerns regulations used to govern the way humans use natural resources so as to maintain such resources for the benefit of future generations. On the other hand, environmental law is trying to find solutions to environmental problems by setting up procedures and guidelines to be followed by decision-makers for purposes relating to the environ-ment.

Liability and compensation are well-understood concepts, and can be applied to effectively manage environmental resources and protect against damage and inap-propriate use. For example, polluters would be held liable for the damage they cause, not just to compensate victims but more importantly to modify their behav-iour so that the damage is averted in the first place. In this way environmental damage is internalized. Common law and statutory law are two mechanisms to achieve it.

Common law has a history of precedents in areas such as strict liability, negli-gence, trespass and nuisance. Issues such as the burden of proof and the costs (and delays) of litigation need to be carefully considered. Common law also extends to the notion of public trust, which both grants and limits the authority of government to control public resources. However, in the area of environmental law, statutory controls have proved to be more effective, particularly where enforced penalties are commensurate with the damage caused.

4.2.2 Environmental planning agencies

Most developed countries have an environmental protection agency that is respon-sible for formulating and implementing policy, including policing and enforce-ment activities. These agencies are given authority via statute. Their jurisdiction may be regional or national and usually covers issues of land-use planning control, pollution and waste control.

4.2.3 Land use and planning control

Land is one of the principle resources available to humankind but is also subject to serious depletion through land degradation, salination, desertification, pollution and soil erosion. Environmentalists argue that if problems of land use are not tackled as a matter of urgency, these will form an increasing threat not just to present activities but also to those that will occur in the future. Economists argue that the most effective way of land management is through the public ownership of land and land-use planning control. Planning is an instrument used to ensure that the proposed use of land will suit the locality and will not interfere with the use of neighbouring land. It is an attempt to choose the best or most suitable use for land, and may lead to the preparation of local and regional environmental plans.

Land-use planning is an important and often controversial environmental issue, particularly as new development extends into remote and previously untouched areas. Botkin and Keller (1995) describe the process of land-use planning as comprising the following steps, all of which should have some form of public participation:

1. Identification and definition of objectives, goals, issues and problems.
2. Collection, analysis and interpretation of data.
3. Development and testing of alternatives.
4. Formulation of land-use plans.
5. Review and adoption of plans.
6. Implementation of plans.
7. Revision and amendment of plans.

4.2.4 Pollution and waste control

Pollution control regulation aims at establishing control on operations and activities that generate pollutants with particular environmental impacts. It generally consists of setting standards and enforcing them by public authorities through the use of licences. Air, water and noise pollution are the main types of pollution that regulations are set up to control.

Pollution control focuses on the point source of pollution, the levels of pollution in the environment and the impact of pollution on humans. As recommended by Field (1997), there are three approaches for pollution control. First, governments may require certain technology or specify production processes to be used to eliminate pollution. Second, governments may set standards for emission and manufacture that all businesses would have to meet. Third, governments may set standards for emissions to be met by polluters in the surrounding environment such as air and water quality standards.

Standards are set by public authorities and are enforced through licensing a plant or product with the right to pollute at a predetermined level. If standards are not met, penalties may result or may even extend to withdrawal of the organization's licence to operate. Field (1997) suggests three types of standards to be used in pollution control:

1. *Ambient standards*. These concern the quality of the surrounding environment such as the ambient quality of air and water. It is normally expressed in terms of average concentration levels over some period of time. Polluters cannot generate pollutants in excess of stated levels. Ambient standards cannot be enforced directly and the standards can only be maintained through certain means of emission control from the source that affect ambient quality levels.
2. *Emission standards*. These are normally expressed in terms of the quantity of emission generated from the pollution sources per unit of time (such as grams per minute, tonnes per week). To some extent setting emission standards does not necessarily guarantee the ambient standard is met. For instance, emission standards can control the emission per source but cannot control the number of

pollution sources. Hence, emission standards are regarded as a type of performance standard because they refer to end results to be achieved by polluters. Polluters are free to choose the best means of achieving these performance criteria.

3. *Technology standards*. Technology standards do not aim at end results; instead, they focus on the types of technology, techniques or practices adopted by potential polluters to control pollution. Technology standards dictate certain decisions and techniques to be used by polluters. These are sometimes described as minimum standards.

The standard of environmental regulation varies greatly from nation to nation. Especially in developing countries, environmental law is often ill-defined. Strict environmental conditions may affect the ability of a nation to be competitive in the international marketplace. International agreements on environmental problems are required to fill in the gaps between national environmental regulations.

4.3 International environmental law

4.3.1 International agreements

Environmental problems commonly have transboundary impacts such as pollution of the air, ocean and the depletion of the ozone layer. They have worldwide impacts rather than being confined within national boundaries. The term 'global commons' refers to those resources that do not belong to a particular country, but are shared across countries. It is very difficult to protect the environment of the global commons since no one nation has absolute authority. They are really the joint concern of all humankind and consequently rely on international participation and co-operation to solve problems. International co-operation is necessary to ensure their effective management.

There are various types of environmental laws that operate at an international level. These include:

1. *Treaties and conventions*. These are agreements made between two (bilateral) or more (multilateral) countries. Some of them are adopted through the United Nations. Countries which sign the treaties or conventions agree to be bound by them. However, the signing of a treaty alone does not make it effective because the international law needs to be ratified by all party nations. This is a lengthy process and only binds those countries that choose to ratify them. Enforcement can be difficult because some countries will not sign if the convention provides for strong action against transgressors.
2. *Declarations*. These are documents adopted by countries but which have no binding effects on countries taking part in the declarations. They are usually significant statements about environmental rights at the international level. They provide the basis for future treaties and conventions.
3. *International court decisions and international arbitration*. The International

Court of Justice (ICJ) is the main court of the United Nations in deciding transnational cases. It consists of 15 judges represented from different countries worldwide. They decide cases at international levels and give opinions on legal questions at the request of the United Nations.

4. *Customary international law*. A principle of customary law has been developed that one sovereign state owes a duty to stop adverse acts to another sovereign state especially in the problem of transboundary pollution. Under customary law a state has a duty to inform other states of activities likely to cause risk to the environment. Countries accept customary law as long as it is consistent with their domestic law.

Field (1997) defines a treaty as an agreement in which all the details have been worked out and included in the signed document. A convention is an agreement in principle in which a general framework is supported but supplemented by future protocols that work out the details.

4.3.2 Implementation problems

Environments that form part of the global commons are equally shared by all nations. It is, therefore, difficult to ensure their protection as no single nation can exercise complete control. Today most environmental problems are found within these global commons and the consequences have international effects. International laws are set up to protect the environment in the global commons, however the impact of international laws is very limited because the governments of individual nations are primarily concerned with national economic growth. They are reluctant to abide by international rules on environmental issues where these are inconsistent with national priorities.

Environmental problems are particularly serious in developing countries where their primary concern centres on economic growth, prevention of starvation and disease control. Environmental protection is often seen as a luxury at this stage of their development. For example, developing countries like India have invested greatly in producing CFC products while the developed countries, which have enjoyed the benefits of CFCs for some years, have announced plans to phase out CFCs to protect the stratospheric ozone layer. It often takes years to negotiate and to persuade the developing countries to agree on certain environmental issues as their attentions are being distracted by more pressing local problems.

International regulations are often difficult to achieve because there is no global government that can impose standards and enforce regulations. Some of the more powerful nations like the United States, Great Britain and Germany may have a virtual veto power over the agreements. Without their participation, international regulations are unlikely to be effective. Even when agreement is reached and a treaty, protocol or convention comes into force, it is difficult to ensure that nations keep to these agreements as no penalties exist for non compliance. The only regulation enforcing mechanism is through trade sanctions against the non-complying nations.

4.3.3 The role of conservation groups

Organizations such as Greenpeace and WWF play a vital role in raising awareness of environmental problems, and through public opinion can encourage governments to develop policies and comply with international agreements. In an indirect fashion, therefore, they can shape and guide environmental law by protesting against bad practices and working with government agencies to implement change.

4.4 Conclusion

The environment of the global commons requires protection and conservation by all nations. It is the hope that both renewable and non-renewable resources can be consumed more efficiently and carefully so that future generations can meet their own needs. Environmental law, even though it cannot save the environment per se, can provide public authorities with power to exercise control over some of the problems introduced through development which can then contribute to the protection and conservation of the environment. Through the implementation of environmental law people will become more aware of the importance of the environment to the social fabric of both present and future generations.

References and bibliography

Barrow, C.J. (1997). *Environmental and Social Impact Assessment: An Introduction.* Arnold.

Bates, G.M. (1995). *Environmental Law in Australia* (4th edition). Butterworths.

Baumol, W.J. and Oates, W.E. (1993). *The Theory of Environmental Policy.* Cambridge University Press.

Beder, S. (1993). *The Nature of Sustainable Development.* Scribe Publications.

Botkin, D. and Keller, E. (1995). *Environmental Science: Earth as a Living Planet.* John Wiley & Sons.

Dragon, A.K. and Jakobsson, K.M. (eds.) (1997). *Sustainability and Global Environmental Policy: New Perspectives.* Edward Elgar Publishing.

Environmental Defender's Office (1992). *Environment and the Law.* CCH Australia Limited.

Erickson, R.L. (1997). *Environmental Law Considerations in Construction Projects.* John Wiley & Sons.

Farrier, D. (1993). *The Environmental Law Handbook.* Redfern Legal Centre Publishing, NSW.

Field, B.C. (1997). *Environmental Economics: An Introduction* (2nd edition). McGraw-Hill.

Fisher, D.E. (1993). *Environmental Law: Text and Materials.* The Law Book Company.

Gilpin, A. (2000). *Environmental Economics: A Critical Overview.* John Wiley & Sons.

Gupta, A. and Asher, M.G. (1998). *Environment and the Developing World: Principles, Policies and Management.* John Wiley & Sons.

Hanley, N., Shogren, J.F. and White, B. (1997). *Environmental Economics in Theory and Practice.* Macmillan.

Hughes, D. (1997). *Environmental Law.* Butterworths.

Hutchinson, A. and Hutchinson, F. (1997). *Environmental Business Management: Sustainable Development in the New Millennium.* McGraw-Hill.

Kenny, M. and Meadowcroft, J. (eds.) (1999). *Planning Sustainability.* Routledge.

Lipman, Z. (1993). *Readings in Environmental Law and Policy: Volumes 1 and 2.* Macquarie University, School of Law, Sydney.

Morgan, R.K. (1998). *Environmental Impact Assessment: A Methodological Perspective*. Kluwer Academic Publishers.

Pearce, D., Markandya, A. and Barbier, E.B. (1990). *Blueprint for a Green Economy*. Earthscan Publications.

Pearce, D.W. and Turner, R.K. (1990). *Economics of Natural Resources and the Environment*. Harvester Wheatsheaf.

Tietenberg, T. (1998). *Environmental Economics and Policy* (2nd edition). Addison-Wesley.

Underdal, A. (ed.) (1997). *The Politics of International Environmental Management*. Kluwer Academic Publishers.

Whitehouse, J. (1991). 'Recent Developments and Trends in Environmental Law'. In *Environmental Law: How the Game has Changed*, pp. 2–9, LAAMS Publications.

Environmental impact assessment

5.1 Introduction

There is significant and increasing awareness of environmental problems, in particular relating to issues such as global warming, stratospheric ozone depletion, biological diversity and population growth. With this increasing awareness comes the realization that the potential impacts of proposed development activities need to be assessed and understood so that appropriate management and control strategies can be adopted. However, projects with significant environmental effects are usually assessed in terms of their impact at a local or regional level.

Environmental Impact Assessment (EIA) has been developed and is used as a means to identify potential damaging effects of proposed developments and is widely accepted as an effective tool of environmental management. EIA was first introduced in the United States in 1969 under the National Environmental Policy Act (NEPA) for land-use planning. Environmental quality was identified as a national priority and the EIA procedure established by NEPA became a worldwide model for comparison.

The definition of EIA is described by Clark (1989) as 'a procedure for encouraging decision-makers to take account of the possible effects of development investments on environmental quality and natural resource productivity and a tool for collecting and assembling the data planners need to make development projects more sustainable and environmentally sound [and] is usually applied in support of policies for a more rational and sustainable use of resources in achieving economic development'. Most countries now have EIA procedures in place and use them to assess projects that are environmentally sensitive (Gilpin, 1995).

5.2 Environmental impact statements

The outcome of EIA is documented in the form of an Environmental Impact Statement (EIS). An EIS is a report that sets out the potential environmental effects flowing from the implementation of a proposed action. The major purpose of an EIS is to present a scientific assessment of the environmental effects of development to decision-makers, regulatory authorities and the community so that the environmental consequences can be forecast. The details of an EIA normally include the environmental impact of the proposed action, adverse environmental impacts that cannot be avoided, alternatives to the proposed action, the relationship between short-term use and the maintenance and enhancement of long-term productivity, and any irreversible and irretrievable commitment of resources (Wathern, 1988). Terms of reference also need to be clearly understood.

The contents of an EIS can vary depending on the project, and there is no international standard for how it might be structured. Gilpin (1995, p. 17) sets out characteristics which an EIS should ideally address. It is interesting to note that one such characteristic is 'the contribution to sustainable development, and the containment of global environmental problems'. The content of an EIS is prepared to suit different types of projects. The detail of EIA content as required by the Council on Environmental Quality (CEQ) for US Federal proposals is provided as an illustration (Wathern, 1988):

1. Summary
2. Statement of purpose and need
3. Alternatives including proposed action
   Discussion of all options considered
   Discussion of 'no-action' option
   Identification of agency-preferred alternative
   Discussion of mitigation measures
4. Affected environment
   Baseline environmental description of area affected by each alternative
5. Environmental consequences
   Environmental impact of each alternative
   Unavoidable effects
   Relationship between local short-term use of environment and enhancement of long-term productivity
   Irreversible and irretrievable commitment of resources
6. List of preparers

The EIS should include discussion of both direct and indirect effects, energy requirements, conservation potential, possible resource depletion, impact on cultural and/or historical resources and urban quality, possible and potential conflicts with local or regional land-use plans, policies and controls, and mitigation measures that might reduce or eliminate potential environmental degradation (Botkin and Keller, 1995).

5.3 Scoping

The EIA process is sometimes criticized because it is a lengthy and costly process. Due to the complex nature of the EIA process and time constraints, EIA will only be carried out on those projects that have significant impact on the environment. The process of creating a list of significant issues to target environmental assessment is called scoping. Scoping is an activity undertaken at an early stage of development and it is designed to focus environmental analysis by examining issues and identifying those that are important. The main purpose of scoping is to notify the public and all relevant organizations and authorities about key aspects of the proposed project.

Scoping can be defined as a very early exercise in an EIA in which an attempt is made to identify the attributes of components of the environment for which there is public concern and upon which the EIA should be focused (Wathern, 1988).

5.4 Impacts identification

After a project has been selected for an EIA process, the next procedure is to identify impacts. A variety of techniques such as checklists, impact matrices, network methods, overlays and simulation models are widely used to assess potential impacts of a project, as follows:

1. Checklists are the simplest approach for rapid assessment of potential impacts, particularly for initial environmental evaluation (Devuyst and Nath, 1993). This technique consists of a simple catalogue of environmental factors that is used to compare against the activities of a proposed development. Checklists are simple to use and give a clear structure to the analysis of important factors, but they must be exhaustive if no serious impact is to be overlooked. However, an exhaustive checklist may become unwieldy and may suppress new initiatives.
2. Impact matrices allow two-dimensional checking of impacts associated with particular types of projects. Leopold et al. (cited in Devuyst and Nath, 1993) were the first to suggest the use of a matrix method for EIA. The developed matrix is particularly useful for EIA as it reflects the impacts that result from the interaction of development activities and the environment. The matrix consists of a list of environmental limits and development characteristics displayed vertically and horizontally to allow checking of one item against the other. Impact matrices commonly represent impacts using a number expressed on a 10-point scale. However, this approach can be misleading as the scores do not have a standardized weighting method. Impact matrices provide a visual summary of the impacts and can help in communicating the results of a study. Impact matrices only can identify direct impacts, and other methods must be used to identify indirect impacts.
3. Network methods are used to account for the indirect impacts of a project. Network methods are flexible and able to show secondary, tertiary and subsequent impacts for each phase of the development process.
4. Overlays are a technique used frequently in environmental planning and landscape architecture. Transparencies are produced showing the spatial distribution and intensity of individual impacts. They, therefore, can be overlain to show total impacts. Overlays are particularly useful with integrated CAD-based software (Wathern, 1988).

5. Simulation models (also known as the EIA method) are simulation-based approaches to EIA. By this method, workshops are used to establish the scope of an appraisal and to identify the key components of environmental systems that may be affected (Wathern, 1988). Computers are used to model real-life situations based on a range of probable variables.

5.5 Impacts evaluation

After impacts have been identified, the next procedure is to evaluate them. There are various evaluation methods in common use including models, experiments, physical representation and valuation. Physical, chemical and biological models can be used to represent the environment. They usually have a mathematical expression so that a mathematical relationship between the cause and effect can be established.

Impacts evaluation can also be based on experiments that are performed in the field or in the laboratory. Impacts evaluation can be presented in the form of physical representations using pictures, photographs, film or three-dimensional models. Finally, valuation methods are used to calculate the cost or benefit of an environmental aspect as a result of an activity.

5.6 Management and control of environmental impacts

As projects typically have negative impacts, they must be managed to ensure that the magnitude of the impacts is within the limits of acceptance, and the project conditions are not violated. The measures that are frequently used comprise mitigation, compensation and monitoring.

Mitigation aims to minimize the impact of a project. The whole project must be analysed to minimize the important impacts at a very early stage. Mitigation measures include those which are inherent to the project such as modification of construction and choice of technology which can remedy or prevent damages. Sometimes no mitigation solutions exist.

Compensation measures are directed mainly towards the public by exposing and indemnifying any bad effects. This may form part of the conditions of approval should the project be constructed. Compensations can be economic, social or sanitary.

Finally, monitoring is the continuous control of environmental variables to ensure that the modifications and actions to be carried out meet set guidelines. Monitoring activities commence at an early stage of a project and may continue through to the dismantling of the installation. They can be classified at four levels, comprising basic monitoring (undertaken prior to development of the project), construction monitoring (undertaken during the construction period), operation monitoring (undertaken during the active life of the project), and finally after-use monitoring (undertaken when the installation is shut down or has finished its operative period). Monitoring adds strength to an EIA and enables enforcement processes to be initiated if violation occurs.

5.7 EIA conditions

It is important that any EIA process minimizes personal bias and maximizes objectivity. The assessment must also be technically and legally defensible. An EIS must provide objective valuation rather than make decisions how to proceed. Gilpin (1995) lists the essential ingredients for the successful undertaking of EIA. These comprise:

1. A legislative basis.
2. A schedule of those works and activities requiring an EIS.
3. Adequate environmental databases.
4. A high level of competence in the physical and social sciences, engineering, planning and environmental management.
5. Integrity in its participants.
6. Active public participation.
7. Avenues of appeal against decisions.
8. Avenues for remedying breaches of the legislation by any person.
9. The ability for any person to require, in certain circumstances, a public inquiry.
10. Competent multi-disciplinary teams for the preparation of EISs and the review of such statements.
11. The ready availability to the public of all relevant documents.
12. The conduct of scoping meetings.
13. The clear allocation of responsibilities for the enforcement of any conditions attached to a statutory development consent.
14. Project and post-project monitoring and auditing.
15. Annual reports on the conduct of the venture.
16. The application of the principle to all policies, plans, programmes and projects likely to have adverse effects on the environment.
17. The clear support of regional environmental plans, and state and national priorities.

Clearly not all these ideal conditions will be achieved on every project, and different countries will place their own values on the importance of the process. But as environmental considerations become germane to development applications then the use of EIA will grow and strengthen.

5.8 Strategic environmental impact assessment

Projects at the EIA level are not developed individually and some projects may relate to other projects, programmes, plans and policies, or their impacts may be linked. Therefore, an assessment of these interactions may need to be carried out. Strategic Environmental Assessment (SEA), also known as programmatic EIA or EIA for policies and programmes, is an assessment of the environmental impacts of more than one project (Devuyst and Nath, 1993).

Some developed countries have been experimenting with SEA, however the structure, procedure and methodology are yet to be defined. The area-wide EIA, developed by the US Department for Housing and Urban Development in the 1970s, is considered to be a type of strategic EIA (Devuyst and Nath, 1993).

5.9 Life cycle environmental impact assessment

Gilpin (1995, p. 14) defines life cycle EIA (LEIA) as 'a procedure for evaluating the environmental impacts of a product, process, or activity, throughout its whole life cycle; a vertical exercise running from cradle to grave'.

Gilpin explains that the main purposes of LEIA are to assess the environmental effects of raw materials of different life cycles, to assess the disposal problem of superseded products, processes and activities, and to evaluate the environmental consequences of alternative processes and design concepts. This process is essentially the same as that undertaken in life-cost studies, except with a focus on physical impacts rather than economics.

5.10 Conclusion

EIA is a procedure for assisting decision-makers to identify and reasonably implement plans, policies and legislation in a form so that environment impacts are properly considered. Even though EIA is neither directly related to the protection of the environment nor provides solutions to environmental problems, it is presently an instrument which is used to aid and improve the decision-making process.

While EIA is a vital process, its main disadvantage is that it is applied to a small range of developments that have significant potential effects as opposed to the majority of developments that have lesser individual yet collectively greater impact.

References and bibliography

Barrow, C.J. (1997). *Environmental and Social Impact Assessment: An Introduction*. Arnold.

Botkin, D. and Keller, E. (1995). *Environmental Science: Earth as a Living Planet*. John Wiley & Sons.

Bowers, J.K. (1997). *Sustainability and Environmental Economics: An Alternative Text*. Longman.

Caimbrone, D.F. (1997). *Environmental Life Cycle Analysis*. Lewis Publishers.

Clark, B.D. (1989). 'Environmental Assessment and Environmental Management', in Proceedings of the 10th International Seminar on Environmental Impact Assessment and Management, University of Aberdeen.

Clayton, A.M.H. and Radcliffe, N.J. (1996). *Sustainability: A Systems Approach*. Earthscan Publications.

Coenen, F.H.J.M., Huitema, D. and O'Toole, L.J. Jr. (eds.) (1998). *Participation and the Quality of Environmental Decision Making*. Kluwer Academic Publishers.

Devuyst, D. and Nath, B. (1993). *Instruments for Implementation: Environmental Management* (volume 3). Vubpress.

Dragon, A.K. and Jakobsson, K.M. (eds.) (1997). *Sustainability and Global Environmental Policy: New Perspectives*. Edward Elgar Publishing.

Fortlage, C.A. (1990). *Environmental Assessment: A Practical Guide*. Billing & Sons.

Gilpin, A. (1995). *Environmental Impact Assessment (EIA): Cutting Edge for the Twenty-First Century*. Cambridge University Press.

Gilpin, A. (2000). *Environmental Economics: A Critical Overview*. John Wiley & Sons.

Gupta, A. and Asher, M.G. (1998). *Environment and the Developing World: Principles, Policies and Management*. John Wiley & Sons.

Hanley, N., Shogren, J.F. and White, B. (1997). *Environmental Economics in Theory and Practice*. Macmillan.

Kaya, Y. and Yokobori, K. (eds.) (1997). *Environment, Energy and Economy: Strategies for Sustainable Development*. United Nations University Press.

Kenny, M. and Meadowcroft, J. (eds.) (1999). *Planning Sustainability*. Routledge.

McCornell, R.L. and Abel, D.C. (1999). *Environmental Issues: Measuring, Analysing and Evaluating*. Prentice Hall.

Morgan, R.K. (1998). *Environmental Impact Assessment: A Methodological Perspective*. Kluwer Academic Publishers.

Nagle, G. and Spencer, K. (1997). *Sustainable Development*. Hodder & Stoughton.

Opschoor, J.B., Button, K.J. and Nijkamp, P. (eds.) (1999). *Environmental Economics and Development*. Edward Elgar Publishing.

Thomas, I. (1996). *Environmental Impact Assessment in Australia: Theory and Practice*. The Federation Press.

Tietenberg, T. (1998). *Environmental Economics and Policy* (2nd edition). Addison-Wesley.

Underdal, A. (ed.) (1997). *The Politics of International Environmental Management*. Kluwer Academic Publishers.

van Pelt, M.J.F. (1993). *Ecological Sustainability and Project Appraisal*. Avebury.

Wathern, P. (1988). *Environmental Impact Assessment: Theory and Practice*. Unwin Hyman.

Environmental policies and strategies

Gerard de Valence

6.1 Introduction

Growing concern over the environment has been matched by the growth in interest in environmental policies. Lobbying and campaigning by environmentalists and community concern over degradation of the natural environment have pressured governments to address the problems associated with externalities more seriously. Economics has an important contribution to make in environmental policy but, despite appearances, there is no single view as to the appropriate form intervention by government should take.

By introducing the idea of externalities, economic models allow us to analyse problems of pollution and resource management. For example, adding all the benefits associated with an activity and all of the costs associated with the activity, including any externalities, determines the socially desirable level of this activity as well as the acceptable amount of pollution arising from it (Devlin and Grafton, 1998, pp. 129–30).

6.2 Political development

In the middle years of the twentieth century, during the spread of industrialization across the world, there was little in the way of environmental awareness and therefore no perceived need for policies to manage the environmental consequences of this rapid expansion in the global economy. This period is now called the 'Golden Age' (1950–70), a time of negligible unemployment, low inflation and boundless prosperity, when living standards rose further and faster than ever before. The emphasis was firmly on the benefits of economic growth and, with few exceptions, little attention was given to externalities associated with the environmental costs of growth (e.g. Yergin, 1991).

Over time, new technologies and new industries greatly increased the productivity of both labour and capital. This increase in productivity has allowed the phenomenal rise in living standards experienced since 1950, financed the rise of the social spending state, and led to the revolution of raised expectations, with each succeeding generation demanding further improvements in their quality of life and lifestyle. The implications of these raised expectations for the future became the focus of much of the concern over the environment (WCED, 1987; Tahvonen and Kuuluvainen, 1993).

However, the 1960s saw the beginnings of greater environmental awareness, sparked by milestone books such as *Silent Spring* (Carson, 1962); the title refers to birds that don't return in spring due to the increasing pollution of the air and waterways and the death of river systems. At the time these concerns were considered politically marginal; on the fringes of social consciousness. Although the evidence was accumulating that environmental issues were, or should be, a matter of concern there were no political organizations or political constituency to promote them (Beavis and Walker, 1983; Hussen, 2000).

The steady growth in environmental awareness over the past three decades has led to the importance placed on environmental policies today, and this process has been driven by a number of shocks to the system – events or information that grabbed people's attention, or more accurately the attention of the media. In particular, public attention was drawn to the fact that species extinction was increasing and biodiversity decreasing, so the first major campaigns were based on wildlife and wilderness preservation (such as whales, seals and Antarctica).

Another concern that became prominent during the 1970s was chemical contamination from the use of pesticides such as DDT, now banned in the western world, due to the recognized health risks. The increasing awareness of the effects of chemicals on community health was reinforced by incidents such the release of dioxin at Seveso (Italy – twice), Bhopal in India and Chernobyl in Russia. Similarly, the effect of CFCs on the ozone layer and the effects of greenhouse gas emissions were major shocks in the 1980s.

During the 1970s these incidents and rising public concern saw the environmental movement emerge, led by organizations such as Greenpeace and Friends of the Earth. During the early 1980s the environment moved towards the political mainstream as these groups became more adept at promoting their cause and the issues began to grab the news headlines. Agendas for the environmental interest groups developed and became increasingly political. At the end of the day, environmental policies are like all other policies, and are created through the political process in one form or another. By this stage the range of policy instruments available to governments had been mapped (Bohm and Russell, 1985; Cropper and Oates, 1992) and the theoretical basis for many of the possible strategies defined (Baumol and Oates, 1988).

Therefore, the 1980s saw environmental protection and management become major issues. On the political front, 'green' parties and independents started contesting elections across the industrialized countries, responding to an emerging mass movement. The agenda during this period became more focused on the political process and competing in elections, particularly in Europe.

It is always more difficult to develop and implement policies when in government, with the compromises this inevitably involves, than to be critical and high-minded in opposition. Also, the jurisdiction of any government is geographically limited, whereas the scope and range of pollution is not (Tietenberg, 1996). Two examples are the acid rain that fell on Germany in the 1970s and 1980s, which was Scandinavian sourced and the subject of lengthy negotiations, and the Japanese subsidies to Chinese coal-fired power generators for scrubbers to reduce power generation emissions carried by wind across the sea to Japan.

Most of the early political success of the environmental movement was at the local or regional government level. Therefore, environmental policies quickly became focused on the three R's, which are: to reduce waste (a design and packaging issue); to reuse as much as possible; and to recycle the rest. The basic aim was, and still is, to reduce the impact of cities on environmental quality. Cities are the major consumers of energy and raw materials and the largest generators of rubbish, waste and pollution. This is where governments, given their limited geographical scope in comparison with the borderless nature of pollution, could quickly have an impact.

The international movement became much more focused on ecologically sustainable development (ESD) culminating in the United Nations Brundtland Report (WCED, 1987). This was extremely important as the definition of ESD used in the Brundtland Report focused on intergenerational equity and the living standards of future generations, which put intergenerational equity at the centre of the policy agenda for the 1990s. Broadly, the Brundtland definition has become the accepted criteria for sustainability and environmental management.

The late 1980s had a much clearer idea of finite resources, and the fact that they would be diminishing in the future as demand increased through population and living standards growth. Also, environmental economics had developed into a well-established body of work with large areas of agreement among practitioners and researchers. The issue of intergenerational equity is compounded by the aspirations of the developing countries to a living standard comparable to that of Europe and North America today. Poverty, growth and the environment are major issues, discussed in Hussen (2000), Kolstad (2000) and Kahn (1998).

A chronology of some of these developments is as follows:

1962	*Silent Spring* published
1972	US Environmental Protection Act passed
1974	First energy crisis
1976	Seveso dioxin release
1979	Second energy crisis
1983	Seveso second contamination
1984	Bhopal gas poisoning
1986	Chernobyl nuclear disaster
1987	Brundtland Report
1989	Exxon Valdez oil spill
1992	Rio Declaration at Second Earth Summit
1997	Kyoto Protocol on global warming
2000	The Hague meeting on global warming

6.3 Environmental issue types

The wide range of environmental issues can be grouped under two broad headings based on the scope of the problems. There are 'green issues' and 'brown issues'. Both were addressed in the Brundtland Report (WCED, 1987) in considerable detail. A different but complementary perspective can be found in Cropper and Oates' (1992) survey of environmental economics.

Green issues include CFCs, greenhouse gases and global warming, wilderness and biodiversity preservation, nuclear testing, ozone and other general pollution issues. They are global in scope because these issues are not solvable at a national level, and any agreements about them are by necessity intergovernmental and therefore very difficult and time-consuming to negotiate. The Montreal Protocol, the Rio Declaration and the Kyoto Protocol on greenhouse gas emissions are the three major international agreements to date.

The debate over implementation of the greenhouse agreement is a good illustration of the difficulty in getting from defining policy objectives to actually implementing policies. Pearce (1991) establishes the case for carbon taxes, Whalley and Wigle (1991) provide a good introduction to the issue of carbon taxes and the international debate over their effects on economic performance, while Winters (1992) has estimated the trade and income effects of greenhouse policies. Despite the agreement on what needs to be done there is currently little prospect of a fully co-ordinated international response, due largely to the uneven distribution of costs in reducing carbon emissions (Weyant, 1993). Emission trading is discussed in Oates (1996), Tietenberg (1996) and Kolstad (2000).

Brown issues are urban in scope rather than global, although to a greater or lesser extent they affect all cities. The single biggest focus is waste disposal, but brown issues also cover the supply of clean air and water, urban amenity, traffic congestion and noise. The Brundtland Report includes the quality of housing and access to services as major urban issues in the developing world.

6.4 Policy as a political process

The environment is now an issue for governments around the world, as public opinion has come to view the degradation of the environment as one of their main concerns for the future (Carraro and Siniscalco, 1993). Also, non-government organizations (NGOs) such as Greenpeace have become influential at the international level, pushing for agreements and protocols between governments on environmental issues (e.g. whaling and greenhouse gases).

Political pluralism is the term used to describe the type of political system found in the OECD countries. It refers to the wide range of opinions debated in an open democratic system, and the interaction of special interest groups and political parties in contributing to the public debate that leads to new policies. Politicians compete for votes in a market place for ideas, and win by proposing, or promising, a better set of policies. Politically, ESD has moved from the fringe to the mainstream of politics and there has been a raising of expectations. Therefore the political costs of not being seen as environmentally friendly is greater than being seen as soft, and this reflects the great change in public perception and community standards.

This is also reflected in the growth of legislation and regulation worldwide in environmental management policies, initially led by the US, Germany and the Scandinavian and Benelux countries. The change from 1980, when there were basically no environmental policies in place, reflects the increasing worldwide concern for the environment. It is no coincidence that democratic countries have implemented environmental policies faster, and have better records in environmental management, than less democratic ones where public opinion plays a lesser role (Gouldson and Roberts, 2000).

6.5 Policy approaches

There are many policy options available to try to minimize the cost of economic activity and degradation of the environment (Baumol and Oates, 1988; Tietenberg, 1990 and 1996). The policy approaches governments are implementing around the world can be classified in a number of ways. One is according to the degree of government intervention required to implement them, which gives three policy options:

1. *Regulation*. This is the administrative fix; usually a complex, expensive system based on enforcement. Like all attempts to regulate markets there are unintended consequences, for example the phasing out of CFCs and the subsequent development of a black market in CFCs in some countries. Favoured in the past, particularly by environmentalists, regulation is less popular with politicians now because of the costs involved in monitoring and the difficulty of enforcement (Russell, 1990).
2. *Economic incentives or market-based solutions*. These aim to change relative prices and thus influence the way in which resources are used. The basic proposition is that the price of a good or service should include all externalities and costs, including environmental costs, and can use quota systems, taxation systems or subsidies to change the way costs are calculated. This approach dates back to the early part of the last century when Pigou (1920) first proposed environmental taxes. The best examples are from the US, where tradeable permits for power generators (Stavins, 1995), auctioning of hunting rights, and differential garbage collection rates based on kerbside sorting have been introduced (Hahn and Stavins, 1992).
3. *Banning a substance*. Usually this has been applied to chemicals such as dioxin or CFCs, regarded as too dangerous to be allowed in the community. Safeguards on certain types of toxic waste are also in this category.

In Australia, for example, virtually all major environmental polices have been based on one form of regulation or another up until now. Direct controls on the quantity of pollutants produced by factories, cars and other sources are common. Most anti-pollution policies are administered by government departments (state and federal), statutory authorities and local councils. The regulatory approach has also been favoured in Western and Eastern Europe, while the US has had more of a mixture of market-based and regulatory policies.

Another way to approach the types of policy is to look at the effect the policy is intended to have on behaviour of producers and consumers. Voluntary programmes are a common approach with limited success rates. They have worked in some cases, particularly voluntary sorting of household glass and paper waste, and more recently the growth of composting. For the corporate sector, competitive pressures tend to work against any one firm acting unilaterally unless there is considerable public pressure. Even where there is a clear policy objective, business and industry do not readily change existing practices.

An example of corporate voluntary action was the response to community pressure by hamburger multi-national McDonald's. In this case, environmentalists in the US were dumping the polystyrene clamshell hamburger containers back onto McDonald's doorstep by the truckload, as their way of protesting about the amount of solid waste the company produced. The reaction to this shows that business responds to community pressure. McDonald's realized that they had a problem (with their image), and engaged a consultancy (the Environmental Defense Fund) that demonstrated that they were indeed producing unnecessarily large amounts of packaging wastes. After an investigation, McDonald's followed the recommendations from the Environmental Defense Fund and reduced solid waste by 70% by changing from clamshells to paper wrappers and reorganizing their pallet system for deliveries. This move gained community credit and was hailed a success by McDonald's. Company representatives then went out to suppliers to find more ways to reduce waste and a quality chain evolved that included nearly 40 000 suppliers in the US alone. Business will respond to community standards by introducing new products or changing practices, especially if this means preserving profits or market share.

The second is direct control, which refers to regulation, enforcement and punishment. This involves controlling the amount of pollution and specifying how activities are to be carried out. Gouldson and Roberts (2000) surveyed over 12 countries and assessed the effectiveness of the regulatory strategy in those countries, and found that it is often poorly enforced.

Economic instruments and taxes are the third and most effective type of policies governments can use to benefit the environment (Tietenberg, 1990). The economic measures for dealing with environmental externalities include tradeable permits, pollution taxes and other economic instruments such as subsidies and deposit-refund schemes, and biomass offsets (e.g. carbon sequestration by forestry plantings).

6.6 Creation of markets for trading property rights

The past ten years have witnessed a dramatic increase in the attention given by policy makers to market-based environmental policy instruments as supplements to the regulations and standards that dominated the previous two decades of environmental law and strategy. One market-based instrument, tradeable property rights applied to power generation emission permits, has been the centre of much of this attention. The idea of using transferable pollution permits to allocate the cost of pollution control across firms or individuals was first developed by Crocker (1966) and Dales (1968). Surveys of the large body of work on tradeable permits can be found in Tietenberg (1980) and Cropper and Oates (1992). Kahn (1998) and Kolstad (2000) show the effectiveness of pollution permits in reaching efficient outcomes, with maximum output for the minimum amount of pollution.

Devlin and Grafton (1998, p. 134) explain the workings as follows:

> Creating tradeable property rights internalises the externality problem by forcing the individual agents to consider the costs of their actions on others. Let's consider the case of pollution. Establishing a tradeable property right over the pollutant itself implies that every unit of pollution costs the firm the price of the right. This statement is true even if the firm does not actually buy the rights because each right it owns has an opportunity cost – the price that it could fetch on the market. As long as the policy maker establishes the appropriate number of rights on the market, the price of the right will reflect social costs and hence the externality associated with a unit of pollution will be internalised (i.e. paid by the firm). We have provided numerous examples of tradeable property rights to deal with pollution, and it has become clear that how these rights are implemented affects critically their ability to solve the problem.

An example is emission taxes in the US, which affect the relative costs of production of power. Hussen (2000) analyses the working of US transferable emission permits for sulphur dioxide emissions, and finds the trading programme has performed successfully. The targeted reductions in emissions have been exceeded at a significantly lower cost than would otherwise have been the case.

Economic measures are not without their problems. The tax and trade approach has some fundamental design issues (Stavins, 1995). The initial distribution of pollution affects the efficiency of a market in tradeable permits (Buchanan, 1969) as does the relative market power of each participant in such a market (Hahn, 1984).

Nevertheless, it is now understood that how the environment is managed and how resources are used depends on the nature of the property rights governing their use. Tietenberg (1996) argues that governments should respond to the changing demand for environmental quality by properly defining property rights systems and improving the efficiency of markets for those property rights. Devlin and Grafton (1998) show how assignment of property rights can be used to control pollution, environmental degradation and use of natural resources.

6.7 Environmental management

In recent years there have been many initiatives in the field of environmental management. Field (1997) argues that any economic system will entail negative environmental impacts if there are no incentives provided to avoid them. Incentive schemes are generally favoured because they involve encouragement rather than punishment. Such incentive schemes can operate at national or regional levels, but may also operate at the level of individual enterprises.

Reinhardt (2000) suggests that environmental management can be seen as a business problem, albeit one with distinctive characteristics and significant public good aspects. Enterprise level incentive schemes are often classified as environmental management systems (EMS). Bragg et al. (1993, p. 4) describes EMS as 'explicit sets of arrangements and processes designed to manage environmental issues and ensure that the organisation's performance goals and objectives are achieved'. There are now international standards for these systems in place.

Environmental auditing is a technique that has emerged as an important component of EMS by playing a significant role in the review and feedback of current performance and the identification of areas of operation that require corrective action. As with the McDonald's example earlier, environmental auditing can help to make improvements to businesses in terms of marketing their image as a corporate citizen, minimizing resource usage and waste, and increasing profit levels.

EMS comprises many facets and is generally regarded as an all-encompassing framework within which specific management and analysis processes can occur. The integration of EMS into the corporate world will significantly advance the environmental movement and this is an area that will undergo enormous development in the years ahead. Environmental auditing will therefore also become a major activity and will generate quantitative data that governments can use to monitor environmental performance.

6.8 Conclusion

There are many possible strategies and policies to deal with pollution, waste and other externalities, and an enormous literature that investigates these. At the end of the day politics is largely about the 'art of the possible', whereas many environmentalists seek ideal solutions to problems. Also, the environmental movement distrusts business and has preferred government regulation to economic incentives and market-based solutions.

Market-based solutions are preferable if they are workable. The cost of maintaining market-based solutions will, however, almost always be less than the cost of monitoring and enforcement. Therefore a consensus for more effective and efficient environmental policies may be the next major development in the international movement towards a sustainable future.

The wide-ranging case studies in Gouldson and Roberts (2000) clearly show that economic and environmental policies are best implemented where there is a high degree of local and regional autonomy, allowing the structure and content of environmental strategies to reflect the social and economic characteristics of a particular region.

With the widespread support that environmental management has in the community, an important element in creating a sustainable future is to direct business research, development and innovation toward better environmental practices, by creating markets where they do not exist (tradeable permits), setting standards or developing new environmental industries. Auditing of performance and the promotion of successful activities in the marketplace will increasingly be an area of involvement for business management.

Individual case studies also show that a planned or programmed implementation of environmental policy, in a coherent and integrated fashion, will achieve better outcomes. Property rights are now considered fundamental for this.

References and bibliography

Baumol, W. and Oates, W. (1988). *The Theory of Environmental Policy*. Cambridge University Press.

Beavis, B. and Walker, M. (1983). 'Random Wastes, Imperfect Monitoring and Environmental Quality Standards'. *Journal of Public Economics*, vol. 21, pp. 377–87.

Best, R. (1996). 'The Building Energy Code of Australia – An Opportunity Missed'. In proceedings of CIB Commission Meetings and Presentation, RMIT, Melbourne, February.

Dohm, P. and Russell, C. (1985). 'Comparative Analysis of Alternative Policy Instruments'. In *Handbook of Natural Resource and Energy Economics* (Volume 1) (A. Kneese and J. Sweeney, eds.), North-Holland.

Bragg, S., Knapp, P. and McLean, R. (1993). *Improving Environmental Performance: A Guide to a Proven and Effective Approach*. Technical Communications (Publishing).

Buchanan, J. (1969). 'External Diseconomies, Corrective Taxes, and Market Structure'. *American Economic Review*, vol. 59, no. 1, pp. 174–77.

Carraro, C. and Siniscalco, D. (1993). 'Strategies for the International Protection of the Environment'. *Journal of Public Economics*, vol. 52, pp. 309–28.

Carson, R. (1962). *Silent Spring*. Macmillan.

Crocker, T.D. (1966). 'The Structuring of Atmospheric Pollution Control Systems'. In *The Economics of Air Pollution* (H. Wolozin, ed.). Norton.

Cropper, M. and Oates, W. (1992). 'Environmental Economics: A Survey'. *Journal of Economic Literature*, vol. 30, no. 2, pp. 675–740.

Dales, J. (1968). *Pollution, Property and Prices*. Toronto University Press.

Devlin, R.A. and Grafton, R.Q. (1998). *Economic Rights and Environmental Wrongs: Property Rights for the Common Good*. Edward Elgar Publishing.

Dragon, A.K. and Jakobsson, K.M. (eds.) (1997). *Sustainability and Global Environmental Policy: New Perspectives*. Edward Elgar Publishing.

Field, B.C. (1997). *Environmental Economics: An Introduction* (2nd edition). McGraw-Hill.

Frosch, R. (1995). 'The Industrial Ecology of the 21st Century'. *Scientific American*, vol. 273, no. 3, pp. 144–47.

Gouldson, A. and Roberts, P. (2000). *Integrating Environment and Economy: Strategies for Local and Regional Government*. Routledge.

Hahn, R. (1984). 'Market Power and Transferable Property Rights'. *Quarterly Journal of Economics*, vol. 99, no. 4, pp. 753–65.

Hahn, R. and Stavins, R. (1992). 'Economic Incentives for Environmental Protection: Integrating Theory and Practice'. *American Economic Review*, vol. 82, pp. 464–68.

Hollinger, D.Y., MacLaren, J.P., Beets, P.N. and Turland, J. (1994). 'Carbon Sequestration by New Zealand's Plantation Forests'. *New Zealand Journal of Forestry Science*, vol. 23, no. 2, pp. 194–208.

Hussen, A.M. (2000). *Principles of Environmental Economics: Economics, Ecology and Public Policy*. Routledge.

Kahn, J.R. (1998). *The Economic Approach to Environmental and Natural Resources* (2nd edition). The Dryden Press.

Kaya, Y. and Yokobori, K. (eds.) (1997). *Environment, Energy and Economy: Strategies for Sustainable Development*. United Nations University Press.

Kolstad, C.D. (2000). *Environmental Economics*. Oxford University Press.

Oates, W.E. (1996). *The Economics of Environmental Regulation*. Edward Elgar Publishing.

Pearce, D. (1991). 'The Role of Carbon Taxes in Adjusting to Global Warming'. *Economic Journal*, vol. 101 (July), pp. 938–48.

Pigou, A. (1920). *The Economics of Welfare*. Macmillan.

Reinhardt, F.L. (2000). *Down to Earth: Applying Business Principles to Environmental Management*. Harvard Business School Press.

Russell, C.S. (1990). 'Monitoring and Enforcement'. In *Public Policies for Environmental Protection* (P. Portney, ed.). Johns Hopkins University Press.

Stavins, R.N. (1995). 'Transaction Costs and Tradeable Permits', *Journal of Environmental Economics and Management*, vol. 29, pp. 133–48.

Tahvonen, O. and Kuuluvainen, J. (1993). 'Economic Growth, Pollution and Renewable Resources'. *Journal of Environmental Economics and Management*, vol. 24, no. 2, pp. 101–18.

Tietenberg, T.H. (1980). 'Transferable Discharge Permits and the Control of Stationary Source Air Pollution: A Survey and Synthesis'. *Land Economics*, vol. 56, pp. 391–416.

Tietenberg, T.H. (1990). 'Economic Instruments for Environmental Regulation'. *Oxford Review of Economic Policy*, vol. 6, no. 1, pp. 17–33.

Tietenberg, T.H. (1996). *Environmental and Natural Resource Economics* (4th edition). Harper Collins College Publishers.

Underdal, A. (ed.) (1997). *The Politics of International Environmental Management*. Kluwer Academic Publishers.

WCED (1987). *Our Common Future* (The Brundtland Report). World Commission on Environment and Development.

Weyant, J. (1993). 'Costs of Reducing Global Carbon Emissions'. *Journal of Economic Perspectives*, vol. 7, no. 4, pp. 27–46.

Whalley, J. and Wigle, R. (1991). 'The International Incidence of Carbon Taxes'. In *Global Warming: Economic Policy Responses* (R. Dornbusch and J. Poterba, eds.), pp. 71–97, MIT Press.

Winters, L.A. (1992). 'The Trade and Welfare Effects of Greenhouse Gas Abatement: A Survey of Empirical Estimates'. In *The Greening of World Trade Issues* (K. Anderson and R. Blackhurst, eds.). Harvester Wheatsheaf.

Yergin, D. (1991). *The Prize*. Simon and Schuster.

Zeckhauser, R. (1981). 'Preferred Policies When There is a Concern'. *Journal of Environmental Economics and Management*, vol. 8, pp. 215–37.

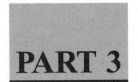

PART 3

Analytical
tools

The majority of analytical approaches employed to evaluate development are monetary-based. While the design of individual projects may need to satisfy development controls imposed by regulatory authorities, the decision as to whether to proceed or not is first one of financial reward. Even where return to the investor is not the primary motivation, overall benefit is usually judged by reference to monetary indicators. An analysis of benefits (inflows) and costs (outflows) over a defined period is used to determine if a proposed project delivers a positive return and if this return is of sufficient magnitude to warrant the trouble and risk of proceeding.

A cost-benefit analysis has use for all development decisions and may be applied to both short-term and long-term proposals. All attributes are monetarized and presented in a cash flow so that the overall gain or loss can be calculated. Determination of return is made more complex in cases of long-term proposals, as cash flow adjustment is necessary to bring all costs and benefits to an equivalent (present) value. This adjustment is known as discounting. Environmental considerations have a similar character to other social concerns and are typically long-term.

The extension of traditional cost-benefit analysis to include social (including environmental) goods and services was a logical progression and an obvious method of integrating both subjective and objective considerations into a single decision model. However, the application of discounting principles to projects with long time horizons has led to criticism of the technique when social issues are involved. Rather than adding weight to the importance of social amenity and environmental quality in the future, discounting devalues it to the extent that after a time period of less than one generation the financial impact of the cash flow is rendered negligible. In the case of environmental goods and services, their value is (if anything) likely to be more highly prized given current trends towards resource depletion, scarcity and reduced access.

This analytical approach also requires the estimation of social issues into monetary terms, which is a difficult task, if not an impossible one. Shadow pricing is often used to assist. For example, the cost of stormwater pollution discharging into a natural watercourse may be estimated as the cost of preventing it in the first place (perhaps by building and periodically cleaning catchment tanks that filter waste) or the cost of cleaning it up once it has occurred (including rectification to beaches, dredging sediment and the like). There is no direct market pricing mechanism for stormwater pollution, so surrogate pricing is used in lieu. The lack of accuracy in estimates of this type, when coupled with the effect of discounting in future years, has generated considerable dissent by conservationists over the suitability of conventional economics to deal with social issues like environment quality. In many cases these issues cannot be monetarized at all, but are left as intangible and outside the main decision criterion.

A favourable cost-benefit analysis is one that demonstrates an acceptable profit and risk allowance within a reasonable time frame. When the principle is applied to public sector projects that have wider social implications, the decision criterion is that collective benefit must exceed collective cost when measured across the community affected by the proposal. Under this rule, some individuals will be losers and others winners, but provided the aggregate benefit of winners outweighs the aggregate loss of losers then the project is considered acceptable.

This concept can be re-expressed for development with significant environmental content. A project is acceptable if it generates capital wealth at the expense of environmental wealth, enabling natural resources to be translated into physical assets of perceived greater value. In this case, society is considered the winner and the environment the loser. This is the main failure of traditional monetary-based decision tools. They assume that any amount of environmental destruction is acceptable provided higher capital wealth is generated, as if at any time in the future environmental wealth can be recreated from accumulated capital.

Yet cost-benefit analysis is still the most popular tool for assessment of development performance. It is purported to identify sustainable projects, but in fact this cannot be guaranteed, as the technique is more concerned with measuring economic progress than environmental conservation. In the case of projects with considerable environmental impact, financial return may be largely theoretical and unrealized as it is used solely for the purpose of choosing between alternative courses of action. Even where costs are included to cover anticipated environmental repair and regeneration, the process does not require that these expenditures ever actually take place.

A combination of financial analysis and legislative control (and enforcement) is therefore necessary if environmental quality is to be protected. Acceptable environmental damage should not be proportional to envisaged benefits but must be limited in absolute terms. The assessment of sustainable development is more complex than financial indicators can model, and requires an evaluation process that is capable of supporting multiple criteria.

The chapters in this part deal with the tools that can be applied to value developments that incorporate social issues like environmental impact. Chapter 7 summarizes the field of environmental economics and looks at methods of assessing the financial implications of essentially non-financial criteria. Chapter 8 explains financial and social cost-benefit analysis, including its advantages and disadvantages. Chapter 9 looks at the difficulties of estimating intangible costs and benefits for inclusion into objective economic models.

Financial decision tools are important and are a part of the quest for sustainable development. However, not everything can be adequately expressed in monetary terms, and therefore other criteria must be brought into the decision process to arrive at a balanced judgement. But it is recognized that if the hard numbers do not add up then the proposal will be rejected, and that other considerations will then be of secondary interest.

Environmental economics

The concept of economics is about the problem of choice and decision. In particular, it concerns economic performance and human behaviour towards scarce resources. The fundamental economic problem is that goods and services are generally scarce. Scarcity exists whenever demand exceeds supply. Nevertheless, goods and services that are free are not considered scarce and hence do not fall into the purview of conventional economics.

Some of the goods and services provided by the environment consist of elements that are scarce in supply. The stock of natural resources such as mineral resources and fossil-based fuels are relatively scarce in supply, while the stock of other environmental goods such as air and water are generally abundant. Under proper management, environmental goods and services should be sufficient to uphold human activities. Unfortunately, due to the growth of population and increased development, the stock of scarce natural resources is threatened with exhaustion. Furthermore, the natural resources that are not scarce are vulnerable to overuse resulting in environmental degradation.

Problems of environmental quality and the increased usage of natural resources and materials are attracting attention from economists, environmentalists, government, community groups and individuals. Many people blame economic growth as the major cause leading to an increase in the environmental loading of production and consumption.

7.2 The importance of environmental economics

Environmental economics focuses on all the different facets of the connection between environmental quality and the economic behaviour of individuals and groups of people. There is a fundamental question of how the economic system shapes economic incentives in ways that lead to environmental degradation as well as improvement. There are major problems in measuring the benefits and costs of environmental quality changes, especially intangible ones. There is a set of complicated macroeconomic questions; for example, the connection between economic growth and environmental impacts and the feedback effects of environmental laws on growth, and there are critical issues of designing environmental policies that are both effective and equitable (Field, 1997).

Environmental economics may be described as the study of environmental problems with the perspective and analytical ideas of economics. Indeed, it is a study of economic decision-making applied to environmental problems by seeking a continuous balance between economic development and environmental quality. The main task is to integrate conventional economic principles with the problems of environmental quality.

Economic activities generate residuals through the process of production and consumption that accumulate in ecosystems resulting in the deterioration of environmental quality. This realization has led to a general concern and scientific interest in the problems of environmental quality. Environmental economics is becoming a useful tool to study environmental problems in a scientific way. It focuses on the quality problems of the physical environment and both environmental degradation and management are identified as key issues.

Environmental economics plays an important role in the identification of possible options for efficient natural resources and environmental management. It is developed to incorporate environmental concerns into the traditional framework of an economic system. Its ultimate aim is to minimize or at least reduce the impacts of human activities on the environment and shift the development process in a more sustainable direction.

7.3 The natural environment

The natural environment has a distinctive feature of time dependence in that its use is spread out over a long period. The supply of resources remains available if it is managed properly. This has obvious implications for future generations and it is the responsibility of the present generation to ensure that this natural stock of resources is not depleted so as to disadvantage the ability of future generations to meet their particular needs.

The natural environment comprises both renewable and non-renewable resources. Broadly speaking, renewable resources exhibit continuous flows through time such as living organisms that are created according to biological processes. Non-renewable resources are natural resources that cannot be replaced once they are consumed. Renewable resources can be living or non-living resources, and non-renewable resources can be further classified as renewable, recyclable and exhaustible (Thampapillai, 1991).

Renewable resources include living biological species in the environment such as micro-organisms, plants and animals. They will be consumed during the process of production but can reoccur naturally in the environment. Renewable resources can also be non-living resources and are often essential elements to maintain life on Earth. They are generally in ample supply and do not incur expenditure in their consumption.

Non-renewable resources diminish upon usage. Once they are used they are gone forever. Non-renewable resources can provide renewable services to support human activities. For instance, fertile lands can support agricultural activities continuously without the need for artificial fertilizers. Nevertheless, if lands are mismanaged and overcropped, desertification may be the result and services may become non-renewable.

Natural resources like tin, copper, gold and iron are non-renewable resources but they can be reused through careful recycling processes. These rely on improvements in recycling technology and human efforts for their conservation. Natural resources like fossil-based fuels are another kind of non-renewable resource that is exhaustible. Their stocks are irreplaceable and their consumption often causes environmental problems like the emission of carbon dioxide, toxic fumes affecting global climate stability and pollution of local ecosystems.

Thampapillai (1991, p. 8) states that 'the natural environment consists of a complex set of resources that range between the categories of renewable and non-renewable ... although some resources or services of the environment are renewable, they could easily become non-renewable in the event of mismanagement'. Indeed, renewable and non-renewable resources require great care in terms of their rate of consumption and human effort in the conservation of environmental quality.

The natural environment is an important component of the economic system, which affects every aspect of humankind now and in the future. Renewable resources should not be consumed at a rate greater than their natural rate of regeneration. Even though non-renewable resources cannot be replaced, they should be conserved and used in a more efficient way. Through technological improvement their conservation can be achieved and their future availability will not be traded-off by the present generation (Pearce and Turner, 1990).

7.4 The role of the natural environment

With regard to the relationship between the environment and economic growth, there are some that advocate a strong focus on economic growth and some that advocate environmental conservation at all costs. Those supporting economic growth argue that it is more important to fix the economy and employment first, then the environment. On the other hand, the environmentalists pronounce the cessation of all economic activities. They claim that humankind is being threatened by the rapidly polluted and degraded environment caused by increased economic activities. Both positions are extreme and somewhere in-between lies the balance that must be struck.

Munasinghe (1993) states that the environment supports the economy in three different aspects. Economic systems rely on nature for raw materials such as mineral resources and timber to maintain the normal functioning of society. If society continues to overuse natural resources a limit on economic growth would necessarily eventuate. In turn, residuals are generated from production and consumption processes and are eventually deposited back in the environment. The environment has the natural capacity to upkeep this balance. Once this balance is disturbed, further deposition of economic residuals will cause depletion of environmental quality. Furthermore, the environment provides amenities to enhance human life and provides essential life-supporting capacity to maintain social well-being.

In reality, the economy is not separate from the environment. There is an interdependence both because economic activities impact on the environment, and because environmental quality impacts on the performance of the economy. The standard of living for humankind is at the interface of environmental conservation and economic growth. On the other hand, the cessation of economic activities will not save the environment. A lack of economic growth will encourage poverty, poor living standards and unemployment, and people may tend to destroy the environment in order to survive. For example, deforestation is more serious in developing countries than in developed countries because people are generally impoverished and they need to clear forests for agriculture.

7.5 The environment as an economic system

The traditional economic system is generally regarded as linear and comprises firms producing consumer and capital goods to create utility or welfare for consumption. Under this linear economic system the role of the natural environment is largely ignored (Pearce and Turner, 1990).

Environmental economists are becoming more concerned about the problems associated with environmental quality and economic growth. They modify the linear economic system by adding the natural environment as part of the process and thus create a cyclical system. The economic function of the natural environment is one of a resource provider. Without the inputs from the natural environment, economic activities simply cannot commence.

This relationship is further complicated because both economic and natural systems generate waste. Natural systems have the capacity of recycling waste into resources. However, economic systems have no such capacity to recycle their own waste. Waste production in economic systems is deposited back into the environment where they can accumulate and degrade natural regenerative processes.

The environment has the capacity to assimilate waste and to convert it back into resources. This assimilative capacity maintains the balance within an ecosystem. Nevertheless, if the deposition of waste exceeds the assimilative capacity of the environment, the economic function of the environment as a waste sink will be impaired. This is how and when environment depletion begins. Furthermore, the excessive deposition of waste will also damage the utility supplied directly by the environment in the form of human enjoyment and comfort.

Recycling is a process to prevent waste from going back into the environment but instead to convert it back to resources via a production process. However not all waste can be recycled. Indeed the majority is not recycled. Waste that is dissipated into an ecosystem will invariably cause environmental problems such as pollution.

Field (1997) recommends three ways of reducing the mass of residuals deposited into the natural environment:

1. Residuals can be reduced by advocating 'zero population growth' (ZPG). Hence the demand on the quantity of goods and services for economic activities is reduced and the impacts on the environment can be more easily controlled.
2. Reduction of economic residuals can be achieved by either technological improvement (such as substituting fossil-based fuels with solar energy) or by shifting high residual output activities to low residual output activities.
3. Residuals can also be reduced by increasing recycling processes. This is particularly important to those recyclable natural resources. Instead of being discharged back into the environment they form an input into other production processes.

An understanding of the relationship between the environment and the economy is fundamental to environmental economics. The ultimate goal is to reduce the depletion of environmental quality caused by the discharge of production and consumption residuals. However, the approaches for reducing residuals are never easy or perfect. For example, reducing population growth cannot guarantee low economic growth, as small populations can have high demand for goods and services. On the other hand, while recycling can turn residuals back into production, the process creates other residuals.

7.6 Economic efficiency

Field (1997) defines benefits as simply outcomes that make people better off. If the environment is cleaned up, benefits are being provided to society. Conversely, if the environment deteriorates, benefits are being removed from society. The idea of benefits is derived from people's willingness to pay or the amount of goods and services that people demand at any particular price, and is built on the assumption that people have preference for a range of goods and services and are able to express their choice.

Benefits are derived from the aggregate demand/willingness to pay curve of defined groups of people. The underlying problems of measuring benefits by aggregate demand/willingness to pay is the difficulty of putting values against environmental assets and measuring the demand on goods and services by people with different income levels. The outcome is also affected by the knowledge of people about the environment, and it is difficult to develop an accurate demand curve if people's knowledge about the environment is weak or ill-informed.

Costs are derived from the use of resources in the production process. The higher demand for a product, the more resources will have to be devoted to its production. Opportunity cost refers to the maximum value of producing an alternative output than a specified one and is measured by the value of inputs used up in production and represented by the marginal cost curve. The true opportunity cost includes all expenditure paid and unpaid in the production process and is determined by the technology utilized in production, price of inputs and the amount of time available to adjust for the new output rate.

Economic efficiency is derived by the relationship between marginal willingness to pay (MWTP) and marginal production costs (MC). The output is efficient where the MWTP curve meets with the MC curve, or in other words maximum net value is obtained by deducting total costs from the total willingness to pay. Therefore, economic efficiency is the level where the benefits enjoyed by the people from the production of a good or service are at the marginal cost incurred in producing it. However, efficiency does not guarantee equality. The benefits may be available to everybody in society or confined only to the rich, and marks the problem of distribution of wealth.

The market is a place where goods and services are traded. The market is at its most efficient level if marginal willingness to pay is equal to marginal costs of production. However, this unique market situation never happens. The market efficiency is affected by the existence of external costs and external benefits such as those concerned with environmental quality. The external costs incurred to correct the damages generated by externalities affect production costs. Where there are substantial differences between market values and social (environmental) values, market failure occurs, and will often require public intervention either to override markets directly or to rearrange things so that they work more efficiently.

7.7 Economic approaches

Economic instruments are not aimed at restricting people to change their actions or behaviour voluntarily. Instead they aim at changing the behaviour of polluters through a focus on monetary return. Environmental law is employed to enable economic instruments to be enforced and to draw people's attention to environmental aspects when making decisions about production, consumption and investment.

7.7.1 Incentive-based approaches

Incentive-based approaches may include taxes, subsidies and deposit-refund systems.

Taxes are prices that are paid for polluting the environment, which provide an incentive for reducing pollution. This is called the polluter pays principle (PPP). Taxes that are most widely used are emission controls and are set on the amount of residuals discharged by polluters over a period of time. This means the polluters are free to determine the way to reduce discharge in order to minimize costs, which may comprise any combination of treatment, internal process changes, changes in inputs, recycling, shifts to less polluting outputs and the like. The idea is to provide an incentive for polluters to find the best way to reduce emissions, rather than having a central authority determine how it should be done.

The introduction of taxes for emission control will eventually raise the cost of production. Nevertheless, polluters cannot just simply pass the taxes on to consumers because competitive pressures will lead the organizations to do whatever they can to minimize their costs. Therefore, as long as there is strong competition in the industry, organizations will respond to taxes and reduce emissions. It is important to set the level of taxes high enough to control emissions. But if the taxes are too high, this will drive organizations out of business and eventually affect the economy as a whole.

Setting the level of taxes is difficult but must be based on measuring the actual discharge against which the tax is to be levied. This may require a permanent monitoring system that measures emissions continuously over time or perhaps will rely on some reasonable judgement of the amount of emissions. This can create arguments between polluters and public authorities over the agreeable amount of emissions to be taxed.

However, this approach enables public authorities to levy a uniform set of taxes so as to achieve specific acceptability standards. If the initial taxes do not reduce the pollution to the preset acceptability standards, the response would simply be to raise the rate of tax. Therefore public authorities are able to estimate the tax level to achieve the target reduction in pollution and accomplish the required standard of environmental quality. A tax approach provides a strong incentive to encourage technological change in pollution control arising from research and development. In fact, under this incentive-based approach, polluters must pay for emissions as well as the abatement costs, while with standards they need to pay only for the abatement costs. Therefore, the potential cost savings from new pollution-control techniques are much larger under the tax approach than under standards.

Subsidies are payments made by the authorities to organizations that pollute below a certain prescribed level. They help to increase incentives for polluters to change their practices or they help them to meet environmental standards. They act as a reward for reducing emissions from a preset level. Subsidies include tax deduction and rebates that encourage environmentally-friendly behaviour. Subsidies can also be in the form of grants for particular programmes and projects or for environmental technology. However, the most difficult part of subsidies on emission reduction is to set the level. To polluters the level should be as high as possible. To public authorities, if the level is too high it will have limited effect on environmental problems. If the level is too low, polluters may find it difficult to achieve.

Deposit-refund systems assign prices to polluting products that are refundable if the products are returned to the manufacturer for recycling purposes. It is a combination of a tax and a subsidy that provides the incentive for people to refrain from disposing of these items in environmentally damaging ways. It helps to reduce litter and to encourage a mentality of reuse.

7.7.2 Environmental management approach

Environmental management creates the right to use environmental resources, or to pollute the environment up to a pre-determined limit, and allows these rights to be traded. Economists argue that environmental degradation is due to the incomplete ownership of rights to use valuable resources. This is the problem that usually occurs in the environment for those resources described as 'global commons'. Since there is no ownership, people tend to overuse or exploit them. The concept of this approach is that if people have rights over the use or pollution of natural environments then they will consider and manage those resources sustainably.

This form of environmental management may include one of the following:

1. *Property rights*. Most of the environmental problems often happen to common free-access resources. Under the circumstance of unspecified ownership of these common resources, individuals and organizations seek to maximize their utilities and as a result natural resources get over-exploited and depleted. The natural environment thus becomes a convenient dumping ground for wastes. Property rights is one of the measures which economists argue will enable the global commons to be better managed. The 'private' ownership of common free-access resources leads to increased intergenerational concern over their protection and conservation. Economists believe that in an economy with well-defined and transferable property rights, individuals and organizations have every incentive to use natural resources as efficiently as possible.

2. *Marketable pollution permits*. This system gives the right to pollute to organizations by issuing permits to emit designated types of waste. The amount of emissions by each holder is in accordance with the number of permits in hand. If the emission exceeds the capacity of the permits, penalties may be the result. These pollution permits are marketable through the mechanism of supply and demand. The main problem with this system is to decide the number of permits to be issued into circulation and the distribution of emission rights at any one time. Through this mechanism authorities are able to control pollution by adjusting the number of permits in the market. As a result, if an organization decides to emit more pollutants it has to purchase more permits from the market, and if an organization reduces emissions it can sell the unused permits for profit. As long as the authorities wish to maintain the preset level of pollution they will keep the same number of permits in the marketplace. If demand for permits is increased, the price will increase proportionally. The organization will tend to invest in pollution control equipment if permit costs are higher than abatement costs. Pearce and Turner (1990) describe three types of permit systems, namely ambient permit systems, emission permit systems and pollution offset systems.

7.8 Conclusion

There is a basic linkage between the economy and the environment. Natural systems have a dual role as a supplier of raw material inputs for the economy and a receptor for production and consumption residuals. The impact of economic processes on environmental quality is at the heart of environmental economics. There is a fundamental need for balance in that over a long period of time all resources taken out of the natural system must eventually end up back in that system in one form or another. Where residuals exceed the assimilative capacity of the environment, damages in the form of pollution will result, and where the damage is significant it can affect economic processes themselves and ultimately the welfare of humankind.

References and bibliography

Barrow, C.J. (1997). *Environmental and Social Impact Assessment: An Introduction*. Arnold.

Beder, S. (1993). *The Nature of Sustainable Development*. Scribe Publications.

Bowers, J.K. (1997). *Sustainability and Environmental Economics: An Alternative Text*. Longman.

Clayton, A.M.H. and Radcliffe, N.J. (1996). *Sustainability: A Systems Approach*. Earthscan Publications.

Coenen, F.H.J.M., Huitema, D. and O'Toole, L.J. Jr. (eds.) (1998). *Participation and the Quality of Environmental Decision Making*. Kluwer Academic Publishers.

Common, M. (1995). *Sustainability and Policy: Limits to Economics*. Cambridge University Press.

Dragon, A.K. and Jakobsson, K.M. (eds.) (1997). *Sustainability and Global Environmental Policy: New Perspectives*. Edward Elgar Publishing.

Field, B.C. (1997). *Environmental Economics: An Introduction* (2nd edition). McGraw-Hill.

Gilpin, A. (2000). *Environmental Economics: A Critical Overview*. John Wiley & Sons.

Gupta, A. and Asher, M.G. (1998). *Environment and the Developing World: Principles, Policies and Management*. John Wiley & Sons.

Hanley, N., Shogren, J.F. and White, B. (1997). *Environmental Economics in Theory and Practice*. Macmillan.

Hutchinson, A. and Hutchinson, F. (1997). *Environmental Business Management; Sustainable Development in the New Millennium*. McGraw-Hill.

Kaya, Y. and Yokobori, K. (eds.) (1997). *Environment, Energy and Economy: Strategies for Sustainable Development*. United Nations University Press.

Kenny, M. and Meadowcroft, J. (eds.) (1999). *Planning Sustainability*. Routledge.

Kolstad, C.D. (2000). *Environmental Economics*. Oxford University Press.

Markandya, A. and Richardson, J. (1994). *The Earthscan Reader in Environmental Economics*. Earthscan Publications.

Munasinghe, M. (1993). *Environmental Economics and Sustainable Development*. The World Bank.

Nagle, G. and Spencer, K. (1997). *Sustainable Development*. Hodder & Stoughton.

Nijkamp, P. (1977). *Theory and Application of Environmental Economics* (volume 1). North-Holland Publishing Company.

Opschoor, J.B., Button, K.J. and Nijkamp, P. (eds.) (1999). *Environmental Economics and Development*. Edward Elgar Publishing.

Pearce, D.W. and Turner, R.K. (1990). *Economics of Natural Resources and the Environment*. Harvester Wheatsheaf.

Thampapillai, D.J. (1991). *Environmental Economics*. Oxford University Press.

Tietenberg, T. (1998). *Environmental Economics and Policy* (2nd edition). Addison-Wesley.

Underdal, A. (ed.) (1997). *The Politics of International Environmental Management*. Kluwer Academic Publishers.

Wills, I. (1997). *Economics and the Environment*. Allen & Unwin.

8

Cost-benefit analysis

8.1 Introduction

Cost-benefit analysis (CBA) is a technique for comparing the flows of expenditure (costs) and the flows of revenue (benefits) as a guide to choosing between alternative investments. CBA can be divided into two main types: economic and social.

Economic cost-benefit analysis is commonly used by the private sector to evaluate any investment project. The costs are those involved in acquisition, construction and operation of the facility while the benefits are those received through rent, sale or other form of earning capacity. Both costs and benefits are relevant to the client or providing authority and are used to indicate the level of profitability that can be expected. This type of analysis is also referred to as 'financial analysis'.

Social cost-benefit analysis is more for use in the public sector and is generally applied to large-scale infrastructure projects The costs and benefits include 'intangibles' that cannot easily be measured in monetary terms and 'externalities' that affect society as a whole. Therefore this type of analysis does not indicate whether or not a project is profitable to the client or providing authority, but rather whether the project contributes to the overall welfare of the community. This type of analysis is also referred to as 'economic appraisal'.

In all other ways economic and social cost-benefit analysis are the same. Cost and benefits are identified over a number of years, the net benefit is calculated for each year and discounted. Net present value is used as the main decision criterion, risk analysis is undertaken to test the sensitivity of the decision and a final recommendation is made. Therefore the discounted cash flow technique is at the heart of the CBA methodology.

8.2 Fundamentals of CBA

CBA is a technique for assessing the return on capital employed in an investment project over its economic life, with a view to prioritizing alternative courses of action that exceed established profitability thresholds. It uses discounted cash flow analysis to make judgements about the timing of cash inflows and outflows and envisaged rates of return.

Most experts agree that discounting is fundamental to the correct evaluation of projects involving differential timing in the payment and receipt of cash. Accounting rate of return and simple payback methods, which do not consider changing time value, are quite inadequate substitutes and may produce misleading advice.

The two most common capital budgeting tools used as selection criterion in CBA are net present value (NPV) and internal rate of return (IRR). Both rely on the existence of costs and benefits over a number of years, and lead to the identification and ranking of projects that are financially acceptable for possible selection.

NPV is the sum of the discounted values of all future cash inflows and outflows. If the NPV is positive at the end of the economic life, then the investment will produce a profit with reference to the discount rate selected. If the NPV is negative, then the investment will produce a loss. Mutually exclusive projects should be selected based on the magnitude of the NPV, provided it is positive. IRR is defined as the discount rate that leads to a NPV of zero. Depending on the cash flow pattern, some projects may exhibit multiple IRR while others may have no rate at all.

Table 8.1 is an example of a discounted cash flow for a hypothetical project having an expected economic life of 10 years. The discount rate used is 10% per annum. The NPV at this rate is $57 992 and the IRR is determined by iteration or trial and error at 20.63%. Figure 8.1 illustrates this graphically. The project can be seen to be acceptable at any discount rate that is less than the IRR. No capitalized reversion at the end of 10 years has been allowed.

8.3 The purpose of social CBA

Social cost-benefit analysis is a comparative tool for evaluating the costs and benefits of projects in order to select the most beneficial investment from a range of options from society's viewpoint (Department of Finance, 1991). Social CBA seeks to measure the net benefits of alternative projects to society as a whole rather than to a particular client or providing authority. It thus takes account of the preferences of individuals in the community by calculation of a single overall figure that indicates the net social benefits of the project.

Social CBA is also a technique that assists decisions about the use of society's scarce resources. It is commonly used for public investment projects where the main objective is not profit-making but to ensure that the total benefits that accrue to society are maximized.

Table 8.1 Discounted cash flow analysis ($)

Year n	Cost C	Benefit B	Net benefit NB=B–C	Discounted value DV=NB(1+d)$^{-n}$
0	120 000	0	(120 000)	(120 000)
1	5 000	32 000	27 000	24 545
2	5 000	32 000	27 000	22 314
3	7 500	39 500	32 000	24 042
4	10 000	47 000	37 000	25 271
5	10 000	47 000	37 000	22 974
6	10 000	35 000	25 000	14 112
7	10 000	35 000	25 000	12 829
8	10 000	35 000	25 000	11 663
9	10 000	35 000	25 000	10 602
10	10 000	35 000	25 000	9 640
			NPV =	57 992

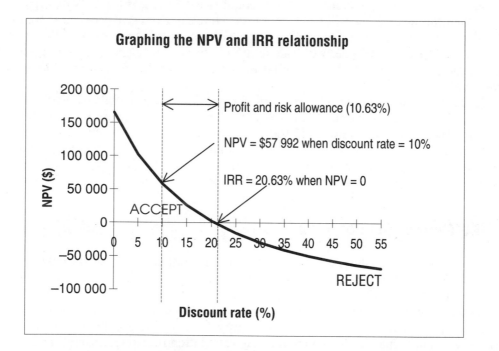

Figure 8.1 Net present value and internal rate of return

The technique is more appropriately used for evaluating large-scale public projects, commonly those contributing to community infrastructure. As large-scale public projects comprise costs and benefits such as time savings, noise, pollution, general amenity and health, they cannot normally be evaluated in terms of pure profit motives. Social CBA has been developed to enable such factors to be translated into monetary terms in order to obtain an objective measure of project worth.

The main area of application has been large infrastructure projects such as power generation, irrigation, airports and road projects.

The technique translates social costs and benefits into monetary terms wherever possible so that they are comparable alongside conventional cash inflows and out-flows. Therefore social CBA is a unique approach for evaluation of costs and benefits which normally cannot be equated together.

Social CBA considers all the factors that reasonably flow from or affect a project in order to identify value. A stone being dropped into a smooth-surfaced pond, creating ripples radiating out to the shore, is a useful analogy for consider-ing the effects of the technique. The analysis must be limited in some way so those effects that are too far removed from the project (secondary effects) are not included in the monetary analysis, although they may be identified in the report. While social CBA is a tool that can aid decision-making it is not a substitute for judgement and experience.

8.4 Aggregate efficiency

Aggregate efficiency (also known as collective utility) is a concept used to describe how goods and services are demanded and fully utilized through the satisfaction of individuals in the community. The satisfaction of an individual is explicitly shown from the users' consumption level of goods and services. Maximum satisfaction can be achieved through the desire of individuals for additional goods and services. Aggregate efficiency is measured by considering the satisfaction or impact of all individuals in the community at large on the decision. Costs and benefits are social rather than private or individual in the sense that they are measured irrespective of the people to whom they occur. Social CBA incorporates the measurement of aggregate efficiency as a funda-mental concept.

Aggregate efficiency is nevertheless a source of some criticism. Although the welfare of the community as a whole is considered, this concept neglects individ-ual distribution of income and wealth. Therefore although society may benefit from a project, certain groups within society may unreasonably suffer.

This problem can be overcome in one of two ways. First, a separate distribu-tional effect analysis can be conducted that examines the impact of the project on specific targeted groups within the community. Alternatively, the costs and bene-fits can be weighted so that the effects of the project can be biased towards those groups in the community that would otherwise be disadvantaged. Normally the first method is employed, as the second becomes ponderous and may still ignore the welfare of small groups of individuals, which although may be highly weighted, may still have little overall impact on the final outcome. Judgement and experience are ultimately required to assess distributional effects. Projects that favour low-income groups in the community are called 'progressive', while those that favour high-income groups are called 'regressive'.

8.5 Allocative efficiency

Allocative efficiency refers to the level of resource usage depicted by the return to society compared to the cost of deployment. To achieve allocative efficiency, the return to society must be maximized and the cost of resource utilization minimized. This can be represented by the calculation of net present value. Allocative efficiency is another fundamental concept of social CBA.

One of the first tasks the analyst faces is to separate the allocative or real effects of a project from its distributional effects. Generally speaking, it is only effects of the former type that alter welfare. Distributional effects are transfers (i.e. some individuals are made better off while others are made worse off) and they do not add or subtract from social welfare. Transfers should be excluded from any social CBA. The danger of double counting is ever present and may occur when measuring costs and benefits that are in fact transfers or which comprise secondary effects.

8.6 Externalities and intangibles

An externality or external effect is an aspect of the project that affects social welfare rather than profit to the providing authority. For the purposes of public project evaluation, an externality occurs when costs and benefits accrue to persons or organizations other than the government or public utility. Externalities are often regarded as 'spillover' effects. They can often be of an environmental nature but can also include matters such as time savings and health.

Externalities can in theory be measured in physical units and valued in accordance to the willingness to pay principle. In practice though, few reliable unit values of externalities are available and difficulties arise in setting appropriate monetary values. Because social CBA takes the community viewpoint, these effects should then be considered within the analysis and a discussion of the impacts included in the project evaluation report.

Intangibles, on the other hand, are simply costs or benefits on which it is difficult or even impossible to place a monetary value. Quite often intangibles are external effects and thus are of concern in the assessment of project performance. The lines of demarcation between externalities and intangibles are often confused, yet the actual classification employed is not usually a problem.

The application of discounting and the translation of social and environmental costs and benefits into monetary terms are two further aspects of the technique that are the subject of much controversy. In particular, the impact of discounting on these subjective social and environmental considerations infers that matters like time savings, the value of human life, lifestyle improvement and environmental protection are linked to the rate at which interest is accrued or lost in the financial market, and this appears inappropriate (Langston, 1994).

Externalities often pose significant problems in their monetary translation, and shadow pricing may become necessary. These costs and benefits can be significant, and unrealistic assessment can easily distort the results. Where externalities are beyond assessment altogether, and hence become intangibles, important aspects of the project can be neglected. Environmental costs and benefits may often be considered as intangible, and clearly this fails to adequately address the sustainability question.

8.7 Methodology

There are several steps in the preparation of a social cost-benefit analysis. The recommended methodology is as follows (Neoh, 1992):

1. *Define the project.* This step defines the scope of the project by an explanation of the rationale and objectives behind the proposal. The beneficiaries of the project should be clearly identified.
2. *Identify the constraints.* All projects have constraints and these should be identified at the outset. Constraints may pertain to financial, physical, legal, administrative, distributional, environmental or other matters. For example, a project that greatly benefits everyone in society may be financially impossible to implement.
3. *Identify the alternatives.* While ideally each alternative that satisfies the objectives for the project should be considered, practically the analysis should be limited to a small number of options (usually not more than five). A 'do-nothing' option should always be identified, either implicitly or explicitly. Careful choice of these options is necessary and to get the best advantage from the process each option should be significantly distinct from the others. Optimization of a particular project in detail is best left until the design process.
4. *Assess the project life and discount rate.* Both project life and discount rate are key variables in the analysis and need to be set for all alternatives. While the assessment of each variable is fraught with problems, viewed collectively the problem is greatly diminished. Public sector projects typically use discount rates of 5% to 10% per annum. At such discount rates the life of the project after 25 to 40 years becomes irrelevant to the outcome. Sensitivity analysis can be employed to test the impact of these variables around the assessed value. Where project lives are unequal between options, assessment must be based on annual equivalents.
5. *Identify the costs and benefits.* The costs and benefits are incremental and accrue both to the providing authority and to all external parties. It is preferable to set out these costs and benefits descriptively in a balance sheet manner. This approach diminishes the chance of double counting, one of the major pitfalls of the technique.
6. *Evaluate the costs and benefits.* All costs and benefits should be converted into monetary terms wherever possible. Some costs and benefits will accrue annually, others intermittently. Externalities will require strategies to be developed to enable their valuation.
7. *Calculate the net present value.* The costs and benefits are formatted into yearly cash flows and the net benefit (benefit minus cost) is calculated for each year

and discounted. The sum of the discounted values is the net present value.

8. *Analyse the risk*. Risk analysis techniques need to be employed to test the sensitivity of each alternative to changes in key variables or assumptions.
9. *Identify distributional effects and other issues*. Where appropriate, the impact of the alternatives of various groups within the community, or various regional areas, needs to be investigated. Environmental or other relevant issues may need to be explored.
10. *Select the best alternative*. The preferred alternative is the one with the highest positive net present value, in the context of (8) and (9) above.
11. *Present the analysis*. The analysis needs to be presented in a form that enables the decision process to be understood by a wide range of individuals. It is important to present the reasons for the recommendation explicitly and with adequate support.

8.8 Advantages and disadvantages

The primary advantages of social cost-benefit analysis over and above those of normal financial analysis are:

1. The ability to identify the projects that maximize the welfare of the community.
2. The ability to objectively assess and quantify the purpose of projects in relation to community needs.
3. The examination of social factors so as to enable projects to be assessed on the basis of benefits other than the pure financial considerations of profit maximization.
4. Recognition of social costs and benefits, such as environmental considerations, which are often treated as having no impact on decisions.
5. Exposure of the basis for decision-making for projects and the opportunity for public criticism.
6. Ability to rank and prioritize limited resources so that the maximum benefit is realized.
7. Decentralized decision-making through the opportunity for community involvement and debate.
8. Less emphasis being placed on personal and intuitive judgements.
9. Wide range of application in terms of the type of projects to which the technique can be applied.
10. Clarity of assumptions.

The primary disadvantages, limitations and pitfalls of social cost-benefit analysis over and above those of normal economic analysis are:

1. Difficulty in measuring social costs and benefits and converting them into monetary terms.
2. Overstatement of the value of social benefits.
3. Complexity.

4. Double counting of costs or benefits through inclusion of secondary effects, for example increased real estate values caused by travel time savings.
5. Time and cost associated with research of social issues and the data needed to price them realistically.
6. Difficulty in defining the limit for the assessment of impacts in the community.
7. Distributional effects not properly considered, thus resulting in equity matters being overlooked.
8. Conflict between social welfare and financial justification.
9. Mistaking of transfers between individuals or groups in the community as costs or benefits attributable to the project.
10. False accuracy due to the use of money as a yardstick for comparison.
11. Self-serving nature of the analysis in that it can be manipulated to justify projects that are politically attractive.
12. Infinite (unmeasurable) values relating to matters such as welfare, environmental quality and human life itself.
13. Unresolved theoretical issues concerning the mechanics of the technique (i.e. the basis of discount rate selection, the treatment of distributional effects, the role of shadow pricing, etc.).

8.9 Specialized forms of cost-benefit analysis

8.9.1 Cost-effectiveness analysis

Cost-effectiveness analysis is generally held to be a type of appraisal that compares the costs and benefits of different initial project options within the framework of a fixed objective (New South Wales Government, 1990a). This objective may be a funding limit or an improvement in the level of services being offered. The focus is on minimizing cost to deliver a specified (fixed) level of benefit. It is therefore a narrower form of cost-benefit analysis, though the terms are often interchanged.

8.9.2 Incidence analysis

Under incidence analysis the overall impacts of the options are disaggregated according to the impact on individual community groups. The disaggregation is commonly undertaken in terms of the income groupings of those affected by a specific development. In other words, it provides information on the distribution of benefits and costs. As such, it provides valuable information to the decision-makers.

8.9.3 Regional impact analysis

Regional impact analysis is the consideration of the effects of a particular project on a specified region. The assessment of regional impact of capital works projects can provide important information to the decision-makers but extreme care should be used in its interpretation. Commonly such analysis relies on regional input/output models to indicate the impact of the project, in terms of income and employment, on the local community and industrial sectors. This form of analysis fails to consider the alternative uses of the resources in the project and associated activities, since regional impact is not equivalent to overall community costs and benefits.

8.9.4 Social impact analysis

Social impact assessment is a technique used to evaluate projects with significant externalities and intangibles that are difficult (if not impossible) to express in monetary terms. A balance sheet approach can be used to weigh up the costs and benefits of a proposal, but rather than expression in discounted dollars a weighted scoring system is commonly adopted. This technique is receiving endorsement, particularly from planners, as an appropriate methodology for evaluation of projects with significant environmental impact.

8.10 Conclusion

CBA is the practical embodiment of discounted cash flow analysis. More frequently the technique is being called upon to evaluate environmentally-sensitive projects, and has come under attack from many conservationists because it can justify investments that cause damage to the environment provided that this damage is outweighed by an increase in capital wealth. In addition, distributional effects are often overlooked or given little weight, as the acceptance criterion is concerned with overall utility improvement and hence does not distinguish whether this benefit is evenly shared.

Use of social CBA is vital as a decision-aiding tool for environmental economics, but caution should be exercised and the final decision to proceed should take into account other issues than merely the maximization of aggregate efficiency.

References and bibliography

Bowers, J.K. (1997). *Sustainability and Environmental Economics: An Alternative Text*. Longman.

Brent, R.J. (1998). *Cost-Benefit Analysis for Developing Countries*. Edward Elgar Publishing.

Clayton, A.M.H. and Radcliffe, N.J. (1996). *Sustainability: A Systems Approach*. Earthscan Publications.

Coenen, F.H.J.M., Huitema, D. and O'Toole, L.J. Jr. (eds.) (1998). *Participation and the Quality of Environmental Decision Making*. Kluwer Academic Publishers.

Common, M. (1995). *Sustainability and Policy: Limits to Economics*. Cambridge University Press.

Department of Finance (1991). *Handbook of CBA*. Australian Government Publishing Service, Canberra.

Department of Quantity Surveying (1993). *Renting v Buying*. University of Technology, Sydney.

Field, B.C. (1997). *Environmental Economics: An Introduction* (2nd edition). McGraw-Hill.

Gilpin, A. (2000). *Environmental Economics: A Critical Overview*. John Wiley & Sons.

Hanley, N., Shogren, J.F. and White, B. (1997). *Environmental Economics in Theory and Practice*. Macmillan.

Kenny, M. and Meadowcroft, J. (eds.) (1999). *Planning Sustainability*. Routledge.

Kolstad, C.D. (2000). *Environmental Economics*. Oxford University Press.

Langston, C.A. (1994). 'The Determination of Equivalent Value in Life-Cost Studies: An Intergenerational Approach'. PhD Dissertation, University of Technology, Sydney.

Layard, R. and Glaister, S. (eds.) (1994). *Cost-Benefit Analysis* (2nd edition). Cambridge University Press.

Nagle, G. and Spencer, K. (1997). *Sustainable Development*. Hodder & Stoughton.

Neoh, Y.L. (1992). 'Cost-Benefit Analysis and the Role of Discounting in Project Evaluation'. B.Build. Dissertation, University of Technology, Sydney.

New South Wales Government (1990a). *Economic Appraisal of Capital Works*. Capital Works Unit, Sydney.

New South Wales Government (1990b). *NSW Government Guidelines for Economic Appraisal*. Capital Works Unit, Sydney.

Opschoor, J.B., Button, K.J. and Nijkamp, P. (eds.) (1999). *Environmental Economics and Development*. Edward Elgar Publishing.

Pearce, D.W. (1988). *Cost-Benefit Analysis*. Macmillan.

Tietenberg, T. (1998). *Environmental Economics and Policy* (2nd edition). Addison-Wesley.

van Pelt, M.J.F. (1993). *Ecological Sustainability and Project Appraisal*. Avebury.

Wills, I. (1997). *Economics and the Environment*. Allen & Unwin.

9

Estimating social costs and benefits

9.1 Introduction

True social costs and benefits change the economic worth of society as a whole. Allocative efficiency is the key principle of cost-benefit analysis. Benefits are valued as the willingness to pay for utilities. Costs are valued as opportunity cost or the foregone return in its best alternative use. Sinden and Thampapillai (1995) use net social benefit to measure economic welfare by subtracting opportunity cost from willingness to pay. The analysis should include all changes in benefits/costs, externalities and unpriced outcomes. It should exclude sunk outcomes, fixed costs, transfer payments, double counting and international outcomes. It should consider taxes and subsidies, government charges, changes in asset values, and secondary benefits and costs. Social costs and benefits may be valued with market prices or without market prices.

9.2 Valuation with market prices

Market price is the most commonly and satisfactorily used approach to value social costs and benefits generated by a project. When the market is competitive, market prices are used to measure the productivity change in terms of input and output (for marginal and non-marginal changes in quantity). Sinden and Thampapillai (1995, p. 83) state that 'a competitive market is characterised by the small market power of participants, mobility of outputs and inputs, unrestrained opportunities for response, homogeneity of goods, and complete knowledge'.

Market demand is the basis of valuing benefits and market supply is the basis for valuing costs. Under marginal change in the production of a good, the benefit of an output (or the cost of an input) is its market price times the change in quantity. In the case of a non-marginal change in production, the benefit of an output is measured by the area below the demand curve and the cost of an input is measured by the area below the supply curve.

The real market is unevenly competitive (distorted) due to government inter-
ventions (taxes, tariffs, subsidies), unemployed labour and foreign ownership. The
shadow prices that adjust the distortions are used to demonstrate the true social
prices. In the case of an increment in availability, the price of outputs (benefit)
includes taxes, and the price of inputs (cost) excludes taxes but includes subsidies.
In the case of displacement of availability, the price of outputs (benefit) excludes
taxes, and the price of inputs (cost) includes taxes but excludes subsidies.

Tariffs and import quotas are used to protect domestic producers when some
inputs are imported. For imported goods, the price should exclude tariffs as it is
merely a transfer of payment between the government and individuals. For domest-
ically produced goods the price is the marginal social cost of production, but when
a project uses previously employed people the labour price is the existing wage
inclusive of taxes. However, when a project uses previously unemployed people,
the labour price is the market wage plus unemployment pay, and when a project
uses previously unemployed people who received unemployment benefits (leisure
value is equal to unemployment benefit) the labour price is the market wage
because unemployment benefits are transfer payments to the unemployed.
However, if a new project creates employment and reduces unemployment benefits the saving in unemployment pay is the real benefit from the project. In the case
of unemployment and underemployment the market wage exceeds the social cost
of labour.

Shadow prices include taxes and royalties paid by a foreign firm because their
benefits accrue to the host country. They exclude net profits to a foreign firm
because they are not benefits that accrue to the host country. Perkins (1994) uses
world prices as the shadow prices where the world market is more competitive than
a domestic market. Distortions in domestic prices are prevented. However, if the
distortions are small, actual market prices are an adequate measure for costs and
benefits.

Market prices are usually available for valuing the output but not the input. Cost
or benefit can be valued by the change in revenue of the output. The benefit of new
road construction to provide access to a factory can be measured through the
change in revenue of the agricultural output of local farmers with and without road
construction. Time can be measured through the employer's willingness to pay for
the change in work output (e.g. time saving increases work output). Wages which
represent the willingness of employers to employ labour is hence used to measure
the value of time (e.g. travel time and work time). This approach is relevant
because it uses actual market data but it has difficulties with estimating the flows
of revenue.

Social costs and benefits can be valued by measuring the change in human productivity, i.e. the change in human capital or earnings. The benefit of water treatment can be measured by the reduced loss in human earnings given that clean
water reduces sickness (and health costs) and maintains human productivity. This
approach can be used to measure environmental amenity and job training that
causes changes in human productivity. It uses real market data but the value of
human life will be understated by merely taking account of human capital.

Changes in cost can affect the benefits of a project. Benefits can be measured
in terms of cost saved by doing something useful (e.g. production cost is saved by

introducing new technology and new machines are used to expedite agricultural output). This is known as the cost-saving method. Alternatively, benefits can be expressed as costs avoided by not doing something detrimental (e.g. soil conservation reduces erosion and salinity, pollution control reduces pollution damage, sound insulation reduces noise pollution, etc.). This is known as the cost-avoiding method or preventive expenditure method. The change-in-cost approach is easy to apply and is relevant using monetary data. However, it is only useful where the quantity of output is fixed because if new technology decreases production costs and increases production output, the cost-saving approach will understate the benefit as it neglects the increase in revenue output.

A social cost or benefit of a project can alternatively be measured by comparing the net present value of a project which generates it and a next best alternative project which does not. The difference between these projects is the opportunity cost of the social cost or social benefit. A project is viable if the social benefit is greater than its opportunity cost, and vice versa. For instance, a new factory may generate more local employment but produce more air pollution. The analyst needs to justify the opportunity cost of a new factory (e.g. potential employment revenue is greater than air pollution) before accepting this proposal. This approach is useful when the social costs and benefits are difficult to measure but does not place proper values on environmental wealth where capital wealth increases are substituted.

The value of a benefit can also be measured by its replacement cost. The minimum value of the benefit is the least cost for replacing it, and vice versa. This approach is useful if the good is replaced precisely but this is not always possible. It can be used to value environmental degradation in terms of resources to replace the environmental services lost. The disadvantage of this approach is that actual replacement cost indicates only the minimum value for the benefit. For instance, timber fencing on a new highway is the minimum value of peace and quiet as it partially replaces the original peace and quiet by reducing noise levels.

9.3 Valuation without market prices

Many costs and benefits that accrue in a project are not exchanged in the market and hence lack market prices. Techniques are required to provide valuations where market prices do not exist.

One such approach measures the social costs and benefits where their quantities are affected by travel times or travel costs. Assumptions are made on the response to visit rates and travel cost or travel times. The social benefit is valued as willingness to pay a cost over and above travel costs to use the facility. The social cost is valued as the welfare loss due to the social disbenefit (e.g. additional travel costs or times). This approach is useful for valuing recreation facilities (such as national parks and scenic areas), road proposals, traffic congestion, historic and cultural areas.

Recreation facilities can be valued by estimating the visitors' willingness to pay to visit the area (i.e. vehicle operating costs, entrance fees and the opportunity cost of their leisure time). The net benefit of a road proposal can be valued by the willingness to pay to use the new road and saved travel costs from the improved road access. Traffic congestion can be valued by estimating the additional travel time costs in the road congestion. Historic and cultural areas can be valued by estimating the willingness to pay to visit these areas and the change in tourism revenue. This approach indicates true value as it is based on actual choices and actual cost data. However, it fails to address various destinations other than the proposal and other factors (further to travel costs or travel times) that affect visit quantity. This approach is only useful if the right assumptions are made and is restricted to facility valuation which are directly related to travel cost or travel time.

Hedonic pricing is used to price a good in a free market where the individuals choose their level of consumption of each good to maximize their own satisfaction. Sinden and Thampapillai (1995) state that the price of a good is the function of the characteristics of the good, the income of the individual and the prices of other goods. The technique has a wide range of applications. It is commonly used for valuing environmental amenity (e.g. scenery, air quality, water quality and land quality) and urban environments by comparing house or land/property prices in proximity to a source of environmental amenity. Perkins (1994, p. 247) states that 'controls are introduced to eliminate differences due to the intrinsic value of the house or land, such as size, age and proximity to shops or other facilities, using large samples and econometric regressions'. It is popular in use because it involves actual prices and actual measurement of characteristics. However, it requires statistical skill to estimate the function. It is more suitable for developed countries where the measurement of environmental amenity comes from sources of environmental degradation and land markets are usually well developed.

Perkins (1994) states that the contingent valuation approach can be used to value a good where no actual market or surrogate market exists, or for goods with utility or non-use value. Non-use or passive values are where the utility is gained without actual usage but by mere knowledge of its existence. This approach estimates the value of a good through the survey responses from the market on maximum willingness to pay for the good. The response to willingness to pay questions can either be the maximum value to pay for a good (such surveys can be done on a face-to-face basis or by telephone), or yes/no answers to a nominated value (such surveys can be done through the mail). The value of a good can be estimated easily by summing and averaging the first type of response. The second type of response gives more precise valuation but requires more complex analytical work.

This approach is useful to measure the value of culturally, historically and spiritually important sites, environmental amenity (e.g. pollution controls, land conservation), other amenities (e.g. fire protection) and human life. Cultural, historical and spiritual sites can be valued through people's willingness to pay to visit them using market prices. Amenities can be valued by estimating the difference in the willingness to pay to visit a site (e.g. recreation site) with or without the amenity. Human life can be valued through the maximum willingness to pay to extend life or reduce the risk of accidents by investments in health, education and transport. Alternatively, human life can be valued through the human capital approach referred to previously.

The major pitfalls of the contingent valuation approach are its hypothetical valuation (based on hypothetical questions and how people value a good), potential biases in the questionnaire and potential untruthful responses. True willingness to pay cannot be revealed if the respondents are biased against a certain type of payment (e.g. dislike payment in the form of income tax). This bias can be avoided by checking whether bids vary significantly with different modes of payment or using neutral types of payment (i.e. donation). The accuracy of willingness to pay is affected by the quantity, quality and sequence of information provided. Therefore, information should be of maximum relevance and identical for each subject. The valuation relates to whether the respondent is asked to pay for an amenity or to accept a bribe to forgo it (Perkins, 1994). Respondents who have no interest in the survey may not be truthful to a nominated value question. This bias can be avoided by pre-surveying the likely value, or if the respondents are informed that they will be paying increased taxes and costs of goods for the amenity.

Willingness to pay for an amenity will be understated if the respondents think they have to pay for it and expect true willingness to pay to come from others. Bohm (1972) overcomes this bias by concealing whether the respondents will be charged for the amenity so that some overstatement of benefits can be neutralized by some understatement of benefits. Valid responses can be determined if the hypothetical bids are checked against some other interview methods, against some other payment modes or against preferences and attitudes, and they should be further checked to determine if the value is consistent with the characteristics of the respondent. True willingness to pay can be obtained if a well-prepared questionnaire is presented carefully to a co-operative group of subjects.

The defensive expenditure technique has been recently developed to value benefits by estimating the willingness to pay for an expenditure to prevent deterioration and maintain the existing utility (i.e. vaccination to babies to prevent sickness, water treatment for safe drinking consumption, etc.).

9.4 The concept of externalities

Often a project will directly or indirectly generate external effects outside its normal operation. These external effects are known as externalities and affect the well-being of other people or businesses. Perkins (1994, p. 239) states that 'when the existence or operation of a project results in a net gain or loss to society but not those who undertake the project, then this category of benefit or cost is defined as an external effect or an externality'.

Externalities can have a positive effect that is beneficial to the greater community, such as may be the case with the hospitality or tourist industry. They can also have a negative effect, perhaps associated with pollution and environmental degradation, such as the effect on the fishing industry of pollution to local waterways.

In practice negative externalities are often ignored if they do not form part of a project's financial cash flow or are not involved in production costs. This is because most negative externalities are experienced only by the local community and not necessarily by those who consume the products or who pay for its production. Therefore, it will not affect the willingness to pay of consumers for the product.

Due to the complex nature of externalities it is difficult to identify their existence and it is also not easy to quantify their value. Measuring the impact of an externality involves putting a monetary value on collective welfare and pricing resources generally available or in common usage. Externalities can be generated during the production, distribution and consumption processes of a project. Externalities created during production processes could include water and air pollution resulting from mining, manufacture and waste treatment. Externalities created during distribution of products could include pollution and traffic congestion. Externalities created during consumption processes could include waste disposal of packaging and health effects such as respiratory diseases caused by passive smoking.

Projects that generate negative externalities do not necessarily compensate third parties who are harmed by their operation. Nevertheless it is important to incorporate externalities in a project's cash flow and reflect the outcomes as part of the overall appraisal. Externalities can be incorporated into the cash flow in a number of ways:

1. Externalities can be internalized during the project's design stage. Hence externalities will become a direct cost or benefit of the project. If the outcome is negative, the designer may change the design to reduce such negative impacts, or prevention measures can be installed such as anti-pollution devices to prevent externalities from happening.
2. In some circumstances the technology to control pollution may not exist or the pollution prevention measures may be too expensive. The developer may choose not to produce at all or may choose to compensate those suffering from the pollution. However, compensation may be difficult to devise and calculate if victims have inadequate property rights. The internalization of externalities through compensation will raise the cost of production of a project and may render it uneconomic.
3. Externalities can be internalized into a project development through government intervention. This is particularly important when legal property rights are unclear. Taxes can be imposed on particular consumer goods that cause large negative externalities to discourage their consumption and internalize the high social cost. An example of this is government taxes on tobacco and alcoholic products.

9.5 Sustainability constraint

The use of a sustainability constraint as part of the evaluation of investment projects is an alternate approach to the incorporation of social costs and benefits, and is closely linked to the choice of a time horizon. Most literature advocates a study period based on either economic life or holding period. There appears to be no universal acceptance of one or other method. Further approaches to the determination of the study period, such as physical life and the analysis of premature obsolescence, are generally not given much regard.

The determination of the study period for an investment project highlights the dilemma between the often opposing objectives of investor interest and social welfare. While it may be conceptually correct to optimize a project so that society achieves the maximum benefit possible, this strategy may also result in no study ever being commissioned since the investor may be departing on a course that ultimately reduces the expected return. Issues of energy conservation and public sector investment suggest social welfare (economic life) is the most appropriate basis for the study period. Issues of profit maximization and private sector investment suggest investor interest (holding period) is preferable. A solution is required that can support both perspectives.

The problem with the use of the holding period is clearly that the interests of future generations may not be well served by the interests of the owner, investor or developer, and hence decisions might be taken that are sub-optimal from society's perspective. The concept of sustainable development therefore appears inconsistent with a study period that focuses on merely the first owner. However, no study period (other than perpetuity) will ensure that the interests of future generations are protected. For example, a project that produces a high return over its economic life may well be a large user of non-renewable energy resources or may generate toxic waste that irreparably damages the environment even after the project ceases to exist.

The introduction of a sustainability constraint can help overcome this problem. This entails debiting the project with the cost of resource depletion or environmental damage resulting from its existence, expressed in real (undiscounted) terms. This cost is converted into an annual equivalent based on the holding period and distributed accordingly in the cash flow forecasts. Development approval would be conditional on this revised analysis showing a positive net present value at the end of the study period. Such action will have the effect of forcing investors to look at ways in which environmental damage can be minimized. The mechanism to ensure that this process occurs lies squarely with government regulation or legislation as there is no incentive to disadvantage projects that would otherwise be profitable if environmental matters were ignored.

The valuation of environmental damage may prove a formidable task. Nevertheless, in many cases it can be based on the cost of repairing the loss or the cost of alternative current technologies substituting for the loss. The sustainability constraint may be positive or negative depending on the nature of the project and its interaction with the environment. Costs and benefits may be estimated on the basis of actual cash flows such as those resulting from rectification work, waste disposal, cleaning up, the sale of by-products and the like, by the determination of shadow prices or alternatively may simply represent a theoretical allowance equivalent to the expected level of risk.

The use of the term 'sustainability constraint' can also be found in Pearce et al. (1989). It is used in that context to describe the balance of an investor's portfolio, so that projects undertaken which result in environmental damage are compensated by other projects within the investor's portfolio which have environmental benefits so that the total social effect is positive. But this is an unenforceable objective and demonstrates a Utopian view of the world. The consideration of environmental issues needs to be more forthright if it is to succeed in practice. Environmental costs and benefits should be balanced against the expected gains in capital wealth so that, if development approval is to be secured, each project must not diminish the total wealth to society, including both capital and environmental aspects.

The inclusion of environmental considerations into economic appraisal will contribute positively towards sustainable development goals. It will enable the study period to focus on the investor's involvement while still protecting the interests of society at large. This approach, however, does not result in the environment being repaired nor does it necessarily limit the profitability of the investment. Rather it is a more subtle incentive to encourage investors and the professional design teams they engage to pursue the development of environmentally-friendly alternatives.

By way of example, assume that an industrial project is proposed as a viable investment opportunity. It has a positive net present value and will return a reasonable profit during the course of its operation. However, the project will cause pollution to the local waterways and will upon the end of its economic life leave the site in an unusable and toxic condition. The theoretical cost of repairing the environmental damage can be estimated and annualized over the study period. In addition to being financially profitable to the investor, the project needs to demonstrate that it will break-even after the cost of environmental damage is considered. Development approval may in the future become conditional upon such an analysis being positive.

A project that does not meet the proposed regulatory requirement of a positive net present value after inclusion of the sustainability constraint would not be granted planning approval. Faced with this situation the investor would naturally look at strategies that would enable approval to be secured. Using the previous example, there are a number of ways in which this might be undertaken:

1. The cost of expected environmental damage could be reduced by making the project less intrusive, perhaps by spending more money initially to incorporate pollution control mechanisms. This is preferable environmentally but will undoubtedly lower the expected return that the project will deliver to the investor, which may be sufficient to cause the project to be dropped. Nevertheless this is the ideal response.
2. The project could be redesigned so that its total cost is lower or that its total benefit is higher. This action, if done legitimately, would be of some advantage in that value for money to the investor should rise. This will probably not result in a reduction in pollution damage but may nevertheless enable the project to become more efficient in its use of energy and save resources through less frequent maintenance and repair.
3. The cost of environmental damage could be under-estimated: a seemingly easy strategy that will enable the project to satisfy approval requirements while not diminishing the real expected return. This is obviously a matter that needs to be regulated by the creation of standard methodologies and rates of estimation combined with independent auditing.
4. The discount rate could be lowered so that future benefits are given more weight. The choice of discount rate based on objective criteria would minimize the possibility of manipulation. A formula for the determination of discount

rates and its rationale can be found in Langston (1994).
5. The study period could be extended so that more benefits could be included to offset the fixed total of the sustainability constraint. However, due to the diminishing exponential effect of the discounting process this action would not normally pose a serious problem.

All of these actions are clearly subject to abuse. Projects could be 'dressed up' so that they meet regulatory guidelines without causing any change to their true character. But the possibilities for falsification are reduced where independent auditing is undertaken and where standard guidelines for the estimation of environmental effects and the determination of discount rates are established.

9.6 Conclusion

Social costs and benefits affect economic welfare. They should be valued in monetary terms regardless of the existence of market prices. Social costs and benefits that have market prices should be measured first. This is because market prices are not usually available for valuing social costs and benefits, especially externalities. There is no perfect technique to accurately estimate social costs and benefits. Most estimates provide only a minimum measure of benefits or costs.

The accuracy of the estimate is further distorted by the potential bias in survey techniques, potential bias in the scope of measurement by valuation techniques and most critically potential bias in revealing true willingness to pay. Nevertheless, the valuation of unpriced costs and benefits is essential to reduce the potential bias of subjective judgement. Sinden and Thampapillai (1995, p. 109) state that 'the valuation and inclusion of unpriced outcomes will improve the estimates of changes in welfare and improve the analysis [. . . and] valuation will also capture the interests of more of the individuals who are affected by the alternatives, and so improve the acceptance of results'.

The valuation of social costs and benefits can be overcome by using relevant valuation techniques, valid assumptions, appropriate surveying techniques and questionnaires, maximum realism and an audience of actual consumers.

References and bibliography

Barrow, C.J. (1997). *Environmental and Social Impact Assessment: An Introduction*. Arnold.
Bohm, P. (1972). 'Estimating Demand for Public Goods'. *European Economic Review*, vol. 3, pp. 111–30.
Bowers, J.K. (1997). *Sustainability and Environmental Economics: An Alternative Text*. Longman.
Brent, R.J. (1998). *Cost-Benefit Analysis for Developing Countries*. Edward Elgar Publishing.
Coker, A. and Richards, C. (1992). *Valuing the Environment: Economic Approaches to Environmental Evaluation*. Belhaven Press.
Common, M. (1995). *Sustainability and Policy: Limits to Economics*. Cambridge University Press.
Department of Finance (1991). *Handbook of Cost-Benefit Analysis*. Australian Government Publishing Service.
Dragon, A.K. and Jakobsson, K.M. (eds.) (1997). *Sustainability and Global Environmental Policy: New Perspectives*. Edward Elgar Publishing.

Goodin, R. (1982). 'Discounting Discounting'. *Journal of Public Policy*, vol. 2, pp. 53–72.

Hanley, N., Shogren, J.F. and White, B. (1997). *Environmental Economics in Theory and Practice*. Macmillan.

Hoevenagel, R. (1994). 'A Comparison of Economic Valuation Methods'. In *Valuing the Environment: Methodological and Measurement Issues* (R. Pethig, ed.), pp. 251–70, Kluwer Academic Publishers.

Hutchinson, A. and Hutchinson, F. (1997). *Environmental Business Management; Sustainable Development in the New Millennium*. McGraw-Hill.

Kaya, Y. and Yokobori, K. (eds.) (1997). *Environment, Energy and Economy: Strategies for Sustainable Development*. United Nations University Press.

Kolstad, C.D. (2000). *Environmental Economics*. Oxford University Press.

Laird, J. (1991). 'Environmental Accounting: Putting a Value on Natural Resources'. *Our Planet*, vol. 3, no. 1, pp. 16–18.

Langston, C.A. (1994). 'The Determination of Equivalent Value in Life-Cost Studies: An Intergenerational Approach'. PhD Dissertation, University of Technology, Sydney.

Layard, R. and Glaister, S. (eds.) (1994). *Cost-Benefit Analysis* (2nd edition). Cambridge University Press.

McCornell, R.L. and Abel, D.C. (1999). *Environmental Issues: Measuring, Analysing and Evaluating*. Prentice Hall.

Morgan, R.K. (1998). *Environmental Impact Assessment: A Methodological Perspective*. Kluwer Academic Publishers.

Opschoor, J.B., Button, K.J. and Nijkamp, P. (eds.) (1999). *Environmental Economics and Development*. Edward Elgar Publishing.

Pearce, D.W., Markandya, A. and Barbier, E.B. (1989). *Blueprint for a Green Economy*. Earthscan Publications.

Pearman, A. (1994). 'The Use of Stated Preference Methods in the Evaluation of Environmental Change'. In *Valuing the Environment: Methodological and Measurement Issues* (R. Pethig, ed.), pp. 229–49, Kluwer Academic Publishers.

Perkins, F. (1994). *Practical Cost Benefit Analysis: Basic Concepts and Applications*. Macmillan.

Price, C. (1993). *Time, Discounting and Value*. Blackwell.

Shechter, M. and Freeman, S. (1994). 'Non-use Value: Reflection on the Definition and Measurement'. In *Valuing the Environment: Methodological and Measurement Issues* (R. Pethig, ed.), pp. 171–94, Kluwer Academic Publishers.

Sinden, J.A. and Thampapillai, D.J. (1995). *Introduction to Benefit-Cost Analysis*. Longman Australia.

Tietenberg, T. (1998). *Environmental Economics and Policy* (2nd edition). Addison-Wesley.

Underdal, A. (ed.) (1997). *The Politics of International Environmental Management*. Kluwer Academic Publishers.

Watson, H. (1994). 'A Preliminary Assessment of the Levels and Costs of Noise on Arterial Roads in Melbourne'. In *Victorian Transport Externalities Study* (volume 2), pp. 1–10, Environment Protection Authority.

Wills, I. (1997). *Economics and the Environment*. Allen & Unwin.

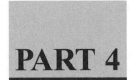

PART 4

Project
feasibility

In pure financial terms, a cost-benefit analysis is extremely useful for determining the contribution that a project is anticipated to make to an investor. It will indicate over a defined time horizon the relative magnitude of the return in comparison to other available investment opportunities. Where discounting is applied, a positive net present value is a normal pre-requisite for acceptance, but the difference between the internal rate of return (or break-even discount rate) and the selected discount rate must be equal to or greater than the minimum expectation of profit contribution and risk contingency.

Risk is a critical factor and requires careful analysis. It is normally expected that projects with high risk levels should also have high profit levels (and hence high internal rates of return), otherwise investors will be reluctant to proceed. Similarly projects of low risk will be able to support lower minimum returns. Risk is analysed essentially by expressing the likelihood of making a suitable profit in the context of a range of possible scenarios for the underlying assumptions.

Project feasibility is often confused with project sustainability. Actually the two concepts have little in common. A project can be feasible because it generates a high benefit-cost ratio (at least greater than one), but may be quite unsustainable in terms of its impact on the environment, its use of energy, and its contribution to community expectations including a whole raft of non-monetary considerations. On the other hand, a project can have minimal impact on its surroundings, be low maintenance, use solar power for heating and cooling, recycle all waste and generally be self-sufficient, yet be totally unfeasible as a financial investment.

Not all projects need make a profit to an investor. This is particularly so for public infrastructure like roads, airports, hospitals and other development that addresses crucial social demand. However, since there are usually limits on available funding, it is important that money is used wisely and to its best advantage. For public sector projects other less tangible considerations are incorporated into the decision-making process so as to demonstrate that the project has merit. Issues such as time saving and productivity, accident reduction, health and well-being, pollution, aesthetics and amenity are set against conventional (tangible) cash flows to determine if there is a theoretical net social gain.

To discover if a project is both feasible and sustainable, which of course is an ideal position, it must first be determined what makes a sustainable project. This is not an easy question to answer and may generate different responses depending on the nature of the development and who is assessing it. Often there is a conflict of interest between profit-seeking investors and amenity-seeking communities that must be managed. Given also that few projects can be truly sustainable, the aim of the exercise is to determine the relative performance of competing alternatives so that a balanced decision can be made.

As a general rule, it is suggested that a project that performs well in terms of sustainability is one that maximizes its contribution to society while minimizing its use of energy (including both embodied and operating energy) and its impact on the environment. To be feasible, a project must have a benefit-cost ratio in excess of one after consideration of all tangible cash flows. The decision process therefore comprises multiple criteria, and an effective balance must be found across all criteria while simultaneously observing minimum standards of compliance to legislated benchmarks. Some flexibility must still be permitted. A project that makes a very strong contribution to society may be allowed to have a higher than normal impact on the environment, or a lower than normal benefit-cost ratio. Each criterion can also be weighted to reflect the specific context of the development. Private sector projects may have more weight ascribed to financial return and energy performance, whereas public sector projects may give more weight to social contribution and environmental impact.

The critical issue is that all projects must be assessed using multiple criteria, and the process of selection must be subject to independent review by authorities responsible for planning approval. Poor performing projects should be denied approval for construction and developers will then be forced to transfer part of their expected profit to overcome the deficiencies. In time projects will be designed to pass these higher approval benchmarks without rework, and constructed facilities will be judged as much by their social contribution as their financial contribution.

The chapters in this part deal with selection criteria and the objective assessment of sustainable development. Chapter 10 focuses primarily on financial feasibility. Chapter 11 identifies the problem of giving appropriate regard to future generations so as not to make decisions that are biased towards short-term objectives. Chapter 12 proposes a method for measuring performance so that competing alternatives can be ranked according to their ability to satisfy key monetary and non-monetary criteria.

In order to make a real difference in the future, the development process must become more sympathetic to sustainability ideals. It is unlikely that this will occur on its own accord across the full range of built infrastructure. If environmental quality is to be protected, more rigorous development controls need to be applied and proper analytical tools must be used that are capable of integrating project feasibility and project sustainability criteria. This must be supported by the progressive introduction of minimum performance standards for completed facilities, a system of self-assessed compliance, and effective enforcement and rectification procedures. The current conflict of interest between investor-centred and society-centred objectives must be resolved through the introduction of enhanced approval processes.

Project selection criteria

10.1 Introduction

Capital budgeting is the process of deciding whether or not to commit resources to a project whose benefits will be spread over a number of future years. The most important factors affecting the evaluation of capital investment are the life of the project and the timing of cash inflows and outflows. Discounted cash flow (DCF) analysis forms the basis for all capital budgeting techniques that take account of the changing value of money. Most experts agree that discounting is fundamental to the correct evaluation of projects involving differential timing in the payment and receipt of cash especially large-scale projects with long time horizons and irregular cash flows.

Correct investment choices will be made if the decision-maker applies the appropriate capital budgeting technique. The selected alternative should yield the highest rate of return and not be lower than the minimum acceptable rate of return (MARR). MARR implies the minimum profitability which a project or project increment must meet to qualify for acceptance. The weighted cost of capital is the theoretical minimum acceptable rate of return. In practice the acceptable standard will be higher. Capital budgeting techniques also must allow all proposals to be ranked according to desirability or profitability.

Capital budgeting techniques that employ DCF comprise net present value, internal rate of return, benefit-cost ratio, savings-investment ratio, domestic resource cost ratio and discounted payback. Conventional capital budgeting techniques that ignore the time value of money like payback period and accounting rate of return are also examined due to their continued high popularity in practice.

10.2 Selection techniques

10.2.1 Net present value

Net present value (NPV) is one of the most useful and commonly employed methods of project selection. It is also known as the net benefits method. NPV is the sum of the discounted present values of net benefits over the life of the project. All future cash inflows and outflows are brought to their discounted present value using an appropriate discount rate. If the NPV is positive at the end of the economic life then the investment will produce a profit with reference to the discount rate selected. If the NPV is negative then the investment will produce a loss.

In the case of independent projects (i.e. projects that will not substitute for each other), the analyst will accept any projects with positive NPVs. In other words, any project which is financially profitable or socially beneficial will be selected. For instance, if the NPV of three independent projects are positive, then under no budget constraint all three projects should be undertaken as they will increase profit or community welfare. But in the case of mutually exclusive projects (i.e. projects that can only be implemented as substitutes for other projects), the analyst will only accept the project with the highest positive NPV. This is because by choosing one mutually exclusive project, the opportunity to do the other will be lost. Projects can be ranked according to the magnitude of their NPV.

In the case of a short-term budget constraint (i.e. the government has a single period budget constraint where it sometimes has a limited investment budget to fund the identified viable projects due to unexpected shortfalls in revenue, unfavourable terms of trade or any other outlays), the government is concerned with the investment costs of the project which can be met from the government revenue in the next few years rather than its operating cost. The government will select a project with the greatest net benefits (receipts minus costs) per unit of investment. A project having the highest NPV is not necessarily the one with the highest returns per unit of investment. However, NPV still can be used for such project selection after some manipulation but the technique can become too cumbersome to handle.

The advantages of the NPV method are its simplicity of use, ease of computation and avoidance of complex conventions about where the costs and benefits are netted out. It is also the only selection criterion that can be used correctly to choose between mutually exclusive projects without further manipulation. On the other hand, the NPV method relies on pre-selection of an appropriate discount rate.

NPV will either be positive or negative. Wright (1990) states that negative NPV implies the project does not meet the target rate, while positive NPV means the rate of return of the project is greater than the target rate and the project is acceptable. However, it does not imply any profitability ranking between projects. To ensure NPV on its own does not mislead the decision-maker, a net benefit investment ratio or internal rate of return should be calculated to provide profitability. The NPV is simple in application but may not be easily comprehended by non-economist administrators or politicians. It is not a suitable method without further manipulation for project selection under a budget constraint as it does not reflect the net benefits per unit of investment. In the case of a long-term budget constraint, the minimum acceptable rate of return will need to be raised to allow fewer projects with a positive NPV otherwise the NPV approach cannot be used for such project selection.

10.2.2 Internal rate of return

The internal rate of return (IRR) is also a commonly used technique of project selection. The IRR of an investment is the discount rate that, if chosen, will lead to a NPV of zero. It does not rely on prior selection of a discount rate. It is commonly established by iteration or trial and error using a financial calculator or computer, or by graphical means. Perkins (1994) states that it would be considered as the maximum interest rate that the project could afford to pay on its fund and still recover all its investment and operating costs. It is called the financial rate of return (FRR) or financial internal rate of return (FIRR) if it is calculated for a financial analysis in which values are measured in market prices. It is termed as economic rate of return (ERR) or economic internal rate of return (EIRR) if it is calculated in the economic appraisal in which values are measured in shadow prices. For instance, the World Bank uses FRR and ERR in its appraisal. The primary purpose of IRR analysis is simply to reflect the relative productiveness of the capital being committed to the project under consideration, not to provide an absolute measure of profitability.

The IRR is compared to the MARR of the investor to determine the economic attractiveness of the investment. In the case of independent projects, all projects with an IRR that exceed the MARR should be accepted (i.e. the project is economic). If IRR is less than MARR, the project should be rejected (i.e. the project is uneconomic). In the case of mutual exclusive projects, the project with the highest IRR is not necessarily the best option. This is because NPV declines when the IRR increases. However, the IRR can be used for selecting between mutually exclusive projects by calculating the IRR of the residual cash flow (the cash flow of the larger project minus the cash flow of the smaller one). In the case of a short-term budget constraint, IRR cannot be used to choose all independent projects whose IRR is greater than MARR. This is because it does not reflect net receipts per unit of project investment which are short in supply. In the case of a long-term budget constraint, MARR should be raised up to the point where those projects with an IRR that exceeds MARR just exhaust the available funds.

The major advantage of IRR is that it is the only project selection technique that incorporates the time value of money without prior selection of a discount rate. This overcomes political problems of selecting different discount rates in different countries and solves technical problems of determining financial and social discount rates. The IRR is commonly used as it is related to the commercial concept of the return on investment and therefore easy for non-economists to understand.

The IRR method implicitly assumes proceeds can be reinvested at the same rate of interest as the rate of return for projects under a budget constraint and it fails to consider the size of the investment, which may incorrectly measure the benefits to be gained. It gives priority to smaller investments with high IRR but low NPV. The calculation of IRR is not possible unless there is at least one negative cash flow period in the project (some projects have no IRR as they do not have any negative cash flow but this is usually not normal). The net benefit stream of some projects has more than one root and hence results in multiple IRR calculation. Graphing IRR for a range of NPVs will ensure that incorrect conclusions about IRR are not inadvertently made.

Marshall (1987) proposes an improvement to IRR, which he calls the adjusted internal rate of return (AIRR). This measure allows returns on reinvestment of net cash flows during the study period at rates different from the rate earned on the original investment. He suggests that this approach provides a more realistic assessment of mutually exclusive projects and does not present the potential problem of yielding multiple IRR solutions. Controversy persists over whether the IRR or AIRR is a more accurate index of an investment's profitability. The former is well known and popular in practice and is considered the more useful decision-making tool when wisely employed. It 'enables the analyst to accept or reject a project on its own earning power' (Montag and Lamp, 1981, p. 55). Ruegg and Marshall (1990) refer to the AIRR measure as overall rate of return (ORR).

Department of Finance (1991) comments that the IRR method is misleading when selecting projects with different lives. A project where benefits accrue soon after the end of the investment period may produce a higher IRR than one where benefits accrue later but in larger amounts. The IRR method involves a quadratic or higher order polynomial equation that necessitates business calculators and computer software programs to solve the complex mathematics.

10.2.3 Benefit-cost ratio and savings-investment ratio

The benefit-cost ratio (BCR) and savings-investment ratio (SIR) are similar to the net present value method. The BCR is the ratio of the sum of the project's discounted benefits to the sum of its discounted investment and operating costs. The SIR is the ratio of the sum of the project's discounted savings to its discounted investment costs. The decision rule is that an investment whose BCR or SIR is greater than one is economic and should be accepted, and if less than one is uneconomic and should be rejected.

In the case of independent projects under a single period budget constraint, the project with the highest BCR/SIR should be accepted up to the point where the budget is exhausted. This is because it gives the highest return per unit of investment when the government investment funds are in short supply. Therefore, the main advantage of the BCR/SIR is that it is the main project selection technique to determine the group of priority projects if there is a single period budget or investment constraint.

BCR/SIR are also termed indices of present value (Bierman and Smidt, 1980) or profitability indices (Wright, 1990) that can be used for profitability ranking between projects by comparison to MARR. If BCR/SIR is less than one, the project should be rejected and if it is greater than or equal to one then the index gives an indication of the relative profitability. BCR/SIR ignores the scale of the investment and leads to acceptance of projects with high BCR/SIR but low NPV.

Perkins (1994) stresses that the BCR of a project will be greater than one when the discount rate is less than IRR. The BCR hence generally should give the same selection results as the NPV method and the IRR method. But the BCR or SIR approach cannot be used to choose mutually exclusive projects and rank projects with a short period budget constraint. This is because a project with the highest BCR or SIR between mutually exclusively projects does not necessary have the highest NPV.

Ruegg and Marshall (1990, p. 54) state that 'the primary application of BCR/SIR is to set funding priorities among projects when there is a limited budget'. The main advantage of the BCR or SIR approach is its ease of comprehension by demonstrating a project's viability in the context of the rise or fall in costs and benefits or savings and investments. BCR/SIR is also sensitive to the way savings and costs are classified in the cash flow as discussed in the NPV approach. It is not a reliable project selection tool and should be used only as a guide on funding priorities for projects under a budget constraint.

10.2.4 Domestic resource cost ratio

The domestic resource cost ratio (DRCR) is used to determine the net benefit to a country of implementing a particular trade-oriented project or trade policy or broader purpose such as determination of a country's comparative advantage (Perkins, 1994). The DRCR is the ratio of the sum of the discounted net local costs to the sum of the discounted net foreign exchange benefits. If the DRCR is less than or equal to the official exchange rate the project should be accepted because it uses less domestic resources measured in local prices to earn one unit of foreign exchange than is the norm for the whole economy. It is also called the internal exchange rate approach and does not involve predetermination of an exchange rate.

The main advantage of the DRCR is that its decision rule can easily be understood by non-economists. It is useful for choosing a potential project that earns foreign exchange in the case of balance of payment constraints. It does not rely on a predetermined shadow exchange rate and hence is an advantage for international agencies. However, it is not suitable for choosing projects in the case of a single period budget constraint and mutually exclusive projects. It neither shows the net receipts per unit of investment for projects under a budget constraint nor demonstrates any mutually exclusive projects that would produce the greatest net benefits to the country. It also requires determination of the true opportunity cost of foreign exchange to the economy for an economic analysis, which is problematic.

10.2.5 Payback period

Payback period (PP) is the most popular of the non-discounting project selection techniques. It is used where there is a shortage in cash and the organization requires the capital investment of a project to be recovered within a specified period. The PP of a project is the period it takes for the stream of net cash flows to equal the initial investment. A project with a PP shorter than the cut-off date specified by the organization is therefore acceptable.

The payback period is still popular due to its ease of computation and comprehension. It is an advantage where the liquidity of a project is vital in assessing its attractiveness. It is a useful decision rule for minor investment projects where cash benefits are fairly regular and also financially too small to justify the time and expense for using a complex but more correct project appraisal technique. It is also useful for selecting projects involving high risks where the period when the capital investment recovered is vital.

However, generally it is a misleading investment decision tool. It ignores the magnitude and timing of all cash inflows and outflows of a project because it neglects the time value of money and only considers cash flows up to the point where capital investment is recovered. It is therefore not a measure of expected return of a project or a project's profitability as all cash flows such as profit after payback period are ignored. It will exclude potential higher profit projects, which have long gestation periods and large cash inflows late in their lives.

10.2.6 Discounted payback

Discounted payback (DP) is the modification of the PP method to take account of the time value of money by discounting cash flows in different time periods. The decision rule is that a project with the shortest discounted payback should be accepted. This method is not a correct measure of project worth as it still ignores the computation of all cash flows arising after the payback period especially where the time profile of benefits differs significantly among alternatives. A project with the shortest DP will not necessary have the highest NPV.

10.2.7 Accounting rate of return

Accounting rate of return (ARR) is also termed as the return on capital employed (Lumby, 1985) or average rate of return (Peirson et al., 1991). The ARR is the ratio of a project's average earnings to its initial investment or to its average investment. Independent projects should be accepted if ARR exceeds MARR. In the case of mutually exclusive projects, a project with the highest ARR should be accepted.

ARR has ease of computation and provides a rate of return that is familiar to management. It is a better technique than PP because it considers all the cash flows over a project's life. However, it is not an appropriate project selection tool because it is ambiguous as to whether initial or average capital is to be employed and how profit is defined. As a result, it provides inconsistent investment decisions. It ignores the financial size of a project and the possibility of differing lives. Most critically it ignores the time value of money and treats all cash flows as real values.

10.3 Risk and uncertainty

The forecasting of future events is normally an integral part of the decision-making process. It also can be the subject of considerable uncertainty and therefore requires cognizance of the level of risk exposure. Evaluation of sustainable development is clearly reliant on appropriate forecasts of future events being made.

The decision-maker is faced with a fundamental choice. Either the uncertainty of the future can be ignored by dealing with only those matters that are known, or estimating tools can be used to make predictions. It is frequently accepted in the literature that it is preferable to plan even when the accuracy of the plan cannot be guaranteed, for otherwise decisions are made in isolation to the environment in which they will ultimately be judged. There are various techniques available to assist in the forecasting of future events. These include extrapolation based on time series or regression, probability distributions, simulations and the like. The choice of technique is a function of the type of forecast being made and the data that is available. But in general, forecasts are made on the basis of historical evidence and hence there is an implicit assumption that the future will be generally consistent.

Forecasting models are abstractions from reality. Models focus on a few characteristics that best represent the real system in a way that is simple enough to understand and manipulate yet similar enough to reality to permit satisfactory results whenever the model is used in decision-making. The test of a model's performance is whether or not it can provide reliable predictions when viewed later using hindsight. Since models must be continually revised, forecasting implies a learning process in which the original model is formulated and the parameters are estimated from empirical data. If the predicted result varies widely from the actual outcome, a critical appraisal must be made to ascertain if the underlying assumptions have been seriously violated. If the assumptions are fallacious, the model should be resimulated to determine whether the previous forecast is accurate given the proper set of assumptions. On the other hand, if the predictions are still at variance with actual results, the model must either be recast or replaced. Thus forecasting is a continual process of technique selection, model construction and forecast evaluation.

Most forecasting techniques will inevitably fail to predict catastrophic events but instead will focus on a range of outcomes that may be reasonably concluded from history. This results in the identification of best, worst and most likely outcomes. If a decision is insensitive to this range then it has a lower level of risk than if it were to easily change. Forecasting adds information to the decision-making process that is a vital part of proper analysis.

The adoption of risk assessment techniques enables the uncertainty of future events to be properly understood. For example, if the discount rate applied to a cost-benefit analysis is likely to vary within a range of values, but at all values the project outcome is favourable, then what is at risk is the level of the benefit not the possibility of a loss. In this case uncertainty remains high but risk is low. Investment decisions are not always so simple, and therefore a number of sophisticated risk assessment techniques are available to quantify the impact of various future scenarios.

Values of many future costs and benefits are usually uncertain. It is risky to simply select an investment that maximizes value without investigating the margin of risk and uncertainty surrounding such single point estimates. Treatment of risk and uncertainty is vital for making a rational investment decision especially if it involves large risk sharing by particular individuals or groups in society. Risk is measurable with known probabilities, while uncertainty is vague without known probabilities.

A risk neutral decision-maker values projects at their expected value. A risk adverse decision-maker values projects at less than their expected value. A risk seeking decision-maker values projects at more than their expected value. A risk neutral approach is applied if it involves small changes in individual wealth, if project risks are independent and if individuals are sharing a large number of risky projects. For instance, public sector projects are borne by taxpayers where an individual's risk exposure is minimal. A risk adverse approach is applied if it involves large risk sharing for particular individuals or groups or there are potential conflicts of interest between managers and owners in the public sector.

Risk is always present when making a decision and does not necessarily create a problem especially where its impact is low. Even if its impact is high, risk can still be accepted by investors because usually the greater the risk the higher the profit they expect to make. Investors must be able to perceive the presence of risks and accurately predict their magnitude and likely impact on the investment. This is accomplished by a systematic and disciplined approach to risk management.

The identification of risk and uncertainty is a central issue for any technique that involves the forecast of future events. Uncertainty (or certainty) refers to a state of knowledge about the variable inputs to an economic analysis, while risk refers to risk exposure or risk attitude (Marshall, 1988). Methods available for quantitatively analysing these matters can be divided into two groups based on whether or not probability theory is involved.

Deterministic approaches to risk analysis include conservative benefit and cost estimating, break-even analysis, sensitivity analysis, risk-adjusted discount rates and certainty equivalent techniques. Probabilistic approaches are more complex and include input estimation using expected values, mean-variance criterion and co-efficient of variation, decision analysis, simulation and mathematical/analytical techniques (Marshall, 1988).

Sensitivity analysis is the most commonly employed risk analysis technique. It involves making minor changes to key variables in order to observe the effect on the originally predicted outcome. Risk and uncertainty can be minimized by demonstrating that the project is not 'sensitive' to such variations. It is dangerous to present results for any investment appraisal without conducting a sensitivity analysis for a range of discount rates.

It is recommended that any risk analysis technique that involves loading the discount rate artificially to include for risk should be avoided. Apart from the implicit assumption that risk is well represented by the compounding interest principle, high discount rates usually result in the selection of low-capital-cost high-operating-cost alternatives that place an unreasonable burden on future generations and are likely to result in the accelerated usage of non-renewable energy resources. This outcome clearly has nothing to do with risk and is hence merely an unfortunate effect.

10.4 Multi-criteria analysis

van Pelt (1993) advocates the employment of multi-criteria analysis (MCA) as an alternate selection technique for projects with significant social costs and benefits. The traditional cost-benefit analysis approach is to take the outcome on one particular criterion as a point of reference, and to adjust it as necessary for other criteria (such as distributional effects). The 'rod of money' is the numeraire, and results are in terms of indicators such as NPV and IRR.

The MCA approach treats all criteria on the same basis without selecting a single criterion (and hence its measurement scale) as a benchmark. Through arithmetical operations, combinations of criteria scores and criteria weights are used to arrive at a ranking of project alternatives.

van Pelt (1993) advocates three primary criteria to be used in the MCA approach, namely efficiency, equity and sustainability. He offers the following solutions:

1. The scores on the efficiency criterion may be of three kinds. If a CBA study allows full coverage and monetization of environmental (and other) effects, comprehensive economic IRRs or NPVs result. If all or the greater part of ecological costs and benefits cannot be monetized, MCA may be applied. An immediate solution is to apply MCA to (a) the outcome of a partial CBA and (b) a group of intangibles, externalities and so on. If MCA is applied, a ranking of alternatives on the efficiency criterion results.
2. The (intratemporal) equity criterion can take several forms. Quantitative indicators may show, for instance, which part of project benefits accrues to target groups. It may also be possible to obtain quantitative indicators involving a weighting mechanism (for instance on the basis of land distribution). MCA may be applied to rank alternatives regarding their overall equity performance.
3. An estimation of the score on the sustainability criterion may take the form of quantitative sustainability indicators. Alternatively, alternatives may be ranked with regard to their sustainability.

The essence of MCA is the use of a numeric score in the various criteria weighted according to their importance. Individual scores are multiplied by weights to arrive at total scores, and this outcome forms the basis of risk analysis and ultimate selection.

10.5 Conclusion

DCF analysis properly forms the basis for all capital budgeting techniques as it takes account of the life and the timing of cash flows. The capital budgeting technique applied should ensure the proposals are profitable enough for comparison, the selected alternative yields the highest rates of return not lower than the minimum acceptable rate of return, and allows all proposals to be ranked according to profitability. Among all capital techniques discussed, NPV is the most highly recommended. The NPV method is the most reliable and appropriate project selection technique because NPV provides an absolute measure of profitability and, can be applied in most project selections. The NPV method can be used for selecting independent projects. It is the only criterion for selecting mutually exclusive projects without further manipulation. It can be used for ranking alternative projects under a budget constraint after some manipulation.

But the NPV method is not without its limitations and therefore it is important that it should not be used in isolation. Evaluation of investments using NPV should include a sensitivity analysis on a range of discount rates to justify project viability. It also should include internal rate of return or net benefit investment ratio to provide a ranking in terms of profitability. Net benefit investment ratio should replace net present value evaluation for assessing projects under a short-term budget constraint.

MCA may offer some advantage to the selection of projects that involve significant social costs and benefits. Further work needs to be undertaken on this approach to develop a rigorous procedural framework.

References and bibliography

Bierman, H. and Smidt, S. (1980). *The Capital Budgeting Decision: Economic Analysis of Investment Projects* (6th edition). Macmillan.

Brent, R.J. (1998). *Cost-Benefit Analysis for Developing Countries*. Edward Elgar Publishing.

Department of Finance (1991). *Handbook of Cost-Benefit Analysis*. Australian Government Publishing Service.

Layard, R. and Glaister, S. (eds.) (1994). *Cost-Benefit Analysis* (2nd edition). Cambridge University Press.

Lumby, S. (1985). *Investment Appraisal* (2nd edition). Van Nostrand Reinhold.

Marshall, H.E. (1987). 'Survey of Selected Methods of Economic Evaluation for Building Decisions'. In proceedings of Fourth International CIB Symposium on Building Economics, Copenhagen, pp. 23–57 (keynotes).

Marshall, H.E. (1988). *Techniques for Treating Uncertainty and Risk in the Economic Evaluation of Building Investments*. US Department of Commerce, Washington.

Montag, G.M. and Lamp, G.E. (1981). *Energy Conservation Evaluation: Rate of Return vs the Present Worth Method of Life Cycle Costing*. ASHRAE Journal, vol. 23, no. 11, pp. 52–55.

Peirson, G., Bird, R., Brown, R. and Howard, P. (1991). *Business Finance* (5th edition). McGraw-Hill.

Perkins, F. (1994). *Practical Cost Benefit Analysis: Basic Concepts and Applications*. Macmillan.

Ruegg, T. and Marshall, H. (1990). *Building Economics: Theory and Practice*. Van Nostrand Reinhold.

van Pelt, M. (1993). *Ecologically Sustainability and Project Appraisal*. Avebury.

Wright, M. (1990). *Using Discounted Cash Flow in Investment Appraisal* (3rd edition). McGraw-Hill.

11

Intergenerational equity

11.1 Introduction

The environmental crisis is of increasing global importance. It is essential to recognize that the environment is affected by human economic activity and yet human economic activity in turn relies totally on the environment. There is by necessity a vital partnership between economics and nature; a balance of immense significance to the long-term prosperity of mankind.

Over the last decade there has been worldwide consensus on the need for sustainable development. Alarming realizations about the rate of depletion of our natural environment caused by the stress of economic growth have resulted in a large proportion of the world's population becoming environmentally conscious. Attitudes towards conservation and pollution are gradually changing, but as the twenty-first century commences this change will, of necessity, need to occur at a faster rate.

Development also implies change, and should by definition lead to an improvement in the quality of life of individuals. Development encompasses not only growth, but general utility and well-being, and involves the transformation of natural resources into productive output. Therefore the environment and the economy necessarily interact. Sustainable development is the balance between economic progress and environmental conservation, recognizing that both are imperative to our future survival.

Sustainable development implies using renewable natural resources in a way which does not eliminate or degrade them or otherwise decrease their usefulness to future generations, and implies using non-renewable natural resources at a rate slow enough to ensure a high probability of an orderly societal transition to new alternatives (Pearce et al., 1989).

The notion of sustainable development places clear emphasis on intergenerational equity. In other words, future generations should not be worse off than present generations and any development should be consistent with such long-term responsibilities. Sustainable development involves an increased emphasis on the value of the natural, built and cultural environments and the recognition that all three are interlinked. Environmental wealth must be considered alongside capital wealth when making investment decisions.

11.2 Equity, justice and distribution

The objectives of society are not clearly distinct from the aims and ambitions of its members, for if people as individuals are better off then they are better off as a group. But a study of economics based on individual behaviour leaves unanswered some of the most pressing problems concerning social policy (Ng, 1983). Intergenerational and intragenerational issues are present and are represented respectively by equity and distributional considerations.

The principle expressed by the sustainability concept itself, that of intergenerational equity, has a number of alternative ethical sources. At its simplest it expounds the Kantian injunction that each generation should do as it would be done by. A more complex interpretation of equity can be derived by asking what distribution of resources would be rational to choose if one were ignorant of the group (or generation) that one belonged to (Tisdell, 1993). In both cases there is a refusal to treat other generations as if they are morally less important than the present one and therefore by implication to discount their lives.

Sustainability is anthropocentric: it wishes to preserve the environment for the benefit of future generations. Intergenerational equity is an expression used to describe the balance in the quality of life experienced by society over time – past, present and future. Where it is achieved succeeding generations inherit a quality of life at least equal in standard to that enjoyed by their forebears. In other words, future generations should not be worse off than present generations. Equity in this context does not mean 'equivalence' but 'fairness'.

But surely intergenerational equity is being achieved. People are healthier, they live longer in bigger, more sophisticated houses, they have access to better technology, they are better informed, fly overseas more often and work shorter hours for more pay. On the other hand, crime levels are increasing, there are more unemployed and homeless people, pollution is worse and our cities are more crowded.

Gyourko (1991) describes quality of life comparisons undertaken in the United States to differentiate between cities. The relative attractiveness of a city is based on what people are willing to pay in order to live there. Matters considered include climate, culture, employment, home prices, wages, crime and pollution, amongst others. Generally these types of comparisons are imprecise and are hard to defend.

Listing the factors that can contribute to our quality of life is a long and difficult task. Quantitative measurement is significantly harder. For example, what importance would be placed on such factors as peace on Earth, human rights or global warming? Even if all the factors were identified, ranked and weighted, it is likely that their assessment would differ wildly among individuals.

The doctrine of sustainable development requires that present development initiatives should not result in a burden for future generations nor diminish their quality of life (Pearce et al., 1989). When viewed at the project level intergenerational equity becomes much more tangible. The selection process for development must ensure that the balance between initial and recurrent costs or benefits is fair and correctly identifies improvements in capital wealth. Projects that result in environmental degradation either explicitly by pollution or implicitly by resource consumption need to also value this correctly to identify the effect on environmental wealth.

The discounting process when used for economic appraisal is intended to express initial and recurrent costs or benefits in equivalent terms so as to enable quantification of capital wealth. Objective valuation of capital and environmental wealth, together with the introduction of appropriate energy and emission controls for all development, will greatly assist in the pursuit of intergenerational equity.

Furthermore, sustainable development implies not simply the creation of wealth and the conservation of resources, but their fair distribution. This is primarily an intragenerational issue. Distributional effects involve utility conflict in that some individuals within society may be better off while others are worse off. Although collectively a positive benefit may occur, the disadvantage accruing to 'minority' groups is of concern. In other words, projects should not solely be based on the principle of majority rules.

In the same manner that equity and justice between generations can be considered by the current valuation of changes in capital and environmental wealth, so too distributional effects can be incorporated in an analysis by the identification of particular social groups and the economic impacts that the project will create. The most effective means to overcome distributional effects which are undesirable is financial compensation. Unlike future generations which are yet unborn, individuals within the present generation can be directly compensated to ensure that fair distribution is achieved.

11.3 Intergenerational equity and discounting

Several researchers have attempted to solve the intergenerational problem by devising specific discounting rules which take explicit account of the expected viewpoint of future generations. These approaches are provided by Price (1993) and repeated below.

Bellinger (1991) models intergenerational discounting through a population which has an even age distribution. Individual members die at a determinate rate and are replaced by new arrivals. Costs and benefits are discounted at one rate over individuals' lifetimes, while discounting between generations is undertaken at a separate intergenerational discount rate. Two special cases to the problem can be identified:

1. Where the intergenerational discount rate is zero the intragenerational rate is derived from some mix of impatience, mortality risk and diminishing marginal utility accruing to the present generation.
2. Where intragenerational and intergenerational discount rates are equal the method gives the conventional net present value (NPV).

Bellinger allows different interpretations according to whether the project evaluated is public or private. For the public project he suggests intragenerational discounting at the individual impatience rate. But if interpersonal comparisons of utility are overlooked, the presumption must be that the only basis for intergenerational discounting is transformation of purchasing power between time periods by reinvestment. This position is endorsed by Freeman (1977), Dobbs (1982), Mishan (1988) and Cropper and Portney (1990).

Such a criterion improves upon ignoring values to future generations entirely. However, impatience does not justify discounting, while potential but unrealized reinvestment is as irrelevant to overlapping generations as it is to those of the present generation.

Discounting rules which purport to embrace intergenerational considerations may be derived from ordinal utility theory. If earlier consumption is preferred over later by every generation, then it is not a surprising deduction that bringing forward any indefinitely protracted programme of rising consumption would also be preferred. Since every generation, looking forward, prefers its own consumption to that of later generations, there is apparently an intergenerational consensus on the propriety of discounting.

However, consistency of intergenerational judgement does not guarantee intergenerational equity. Broome (1992) refers to this consistency as intergenerational neutrality, and distinguishes it from intergenerational impartiality which gives no greater weight to early consumption. Broome shows that evidence indicating that impartiality is impossible results from an arbitrary (and unconvincing) definition of the characteristics of preference.

Consistent preference for earliness throughout generations is actually an artefact of the implicit rule that choices may only be made looking forward: future generations are not allowed regrets about previous generations' decisions. Future generations can be disenfranchised just as effectively by restricting the options they are allowed to vote on as by precluding them from voting at all. Once preferences about consumption prior to one's own generation are allowed, the consensus on discounting is disputable. Even with no backward preferences and an assumption of continuously increasing income from generation to generation, only a strong taste for egalitarian distribution assures discounting of majority support (Brown, 1987).

Where short-term benefits of a project are achieved at a small but indefinitely prolonged cost, the majority of an intergenerational constituency would, in the absence of compensation, oppose the project (Page, 1977). This is not the result given by discounting. Dobbs (1982) objects that a government which is unprepared to compensate future generations is unlikely to be moved by a hypothetical majority of unborn voters.

Whatever the merits of the above rules, they concern only the existence and generational distribution of ordinal time preference, not the intensity of cardinal time preference. Nijkamp and Rouwendal (1988) attempt to merge conventional discounting with a perspective that includes the viewpoint of future generations. The NPV of a long-term project is recalculated for each time (or generation) period, based only on the costs and benefits still to be received.

While Nijkamp and Rouwendal do endorse a system of declining weights in successive periods, they suggest that NPV at time zero should receive uniquely heavy weight. But the formulae they offer bear no relationship to the obvious weighting factor – the number of individuals in the population during each subsequent period. If these population-sized weights are used, the result becomes much more stable to the varying length of generation period. Nonetheless, late cash flows continue to be multiple-counted. Nijkamp and Rouwendal's method increases the importance of future generations' viewpoint. But the means of defining weights is unclear, and the multiple-counting of late revenues and costs favours delayed benefit, which has no obvious justification.

Kula's modified discounting method (Kula, 1981 and 1988) is designed for public projects whose benefits are necessarily distributed across generations. The essence of Kula's philosophy is that individuals have a right, under the terms of consumer sovereignty, to discount within their own lifetimes, but no right to discount utility accruing to future generations. The basis for individual discounting is the product of factors due to mortality risk and to diminishing marginal utility. Over the life-span of individuals alive when an investment begins, the benefits accruing to them are discounted, say at 5% per annum. For new members of the population the discount factor as they join is one, but it declines at 5% each year through their lives.

Kula's method appears to achieve the difficult compromise between intergenerational equity and efficiency as defined by consumers' sovereignty. It also avoids intuitively offensive conclusions such as a perpetual source of utility has finite value. To the extent that it reduces the severity of discounting it favours future generations. However, the method has inconsistencies of three kinds:

1. In practice, the choice of investment depends on the point in time from which it is viewed. Inconsistencies arise due to the changing ratio between discount factors for pairs of dates as those dates become closer. The consequence is that investments tend to become less attractive as their date of initiation approaches.
2. The basis of discounting does not match the structure of the discount factors. Mortality risk does not remain constant through life and yet it appears as a constant in individuals' discount functions. More seriously, mortality in a discount function implies a probabilistic view of death: in the method, however, people die deterministically at the end of the mean life-span. A consequent quirk is that, as expectation of life declines, the mortality discount increases, but long-term discount factors devalue the future less (Price, 1989). On the other hand, diminishing marginal utility (derived by Kula from the general increase of affluence in society) clearly affects the value of consumption by children who are more affluent than their parents were at a given age, just as it affects the value of the parents' own consumption. Yet it does not appear in intergenerational discounting.
3. The claim of equity is inconsistent with the factors actually derived. The mean weight on consumption of older individuals is greater through their remaining life-span than that on consumption of younger ones. A more insidious inequity is that discount factors do not differ much from conventional ones over the first few years: thus the present generation can justify making only small sacrifices for the good of the next generation, yet expect the next

generation to make much larger ones for the generation following (Price, 1989). This lack of required present action is, in practice, a norm that will be handed down the generations, each generation postponing the obligation to make sacrifices.

The inconsistencies in Kula's approach are best evaded by not using the method. To the extent that it is based on constant probability of mortality it is irrelevant to social discounting and inaccurate for individual discounting. To the extent that it is based on diminishing marginal utility, it fails to justify its implicit assumption that diminishing marginal utility will not operate between the present and the time of birth of future generations.

11.4 Affordability and living standards

A better approach to the measurement of intergenerational equity change is to explore the relationship between future prosperity and living standards. If future generations are expected to be better off than the present generation it can be concluded that living standards improve. Living standards and quality of life are generally regarded as synonymous. Intergenerational equity is about ensuring that living standards do not deteriorate over time as a result of current action.

The common difficulty found in measuring changes in living standards is due to the imponderables of social issues pertaining to the quality of life. Working hours and conditions, improvements in technology levels for manufactured goods and services, greater variety of choice in purchase, attainment of lifestyle objectives, law and order and environmental issues are just a few of the facets of living standards that will affect and defy meaningful assessment. Living standards are judged with reference to the individual. The performance of governments and corporations is reflected by the benefit they ultimately provide to their citizens and shareholders. Final analysis must be a function of the perception of individuals in comparison to the standards being achieved by others. Yet there remains no single definition of living standards upon which to base its measurement (Zwartz and Marcus, 1989).

Manning (1984, p. 2) states that the standard of living is a subjective concept, and as such there can be no objective way of measuring its cost. Nevertheless, while 'there are those that are content to leave the standard of living as non-observable, [. . . others] claim that the cost of a constant standard of living can be inferred by the application of an estimating technique to data on household expenditure patterns'. Gilchrist (1986, p. 50) believes that the 'only way households will be able to maintain their standard of living is by changing their spending habits and more strictly controlling their money'.

Living standards therefore form a suitable methodology for the measurement of affordability. A comprehensive review of past research into the measurement of living standards is provided by Brownlee (1990). Four main approaches adopted by researchers in various parts of the world are identified and categorized as follows:

1. *Income and expenditure approach.* The measurement of living standards using income and expenditure patterns has formed a main stream of research effort. Real disposable income, if increasing, is considered to lead to corresponding increases in living standards. Yet others conclude that looking at income alone does not account for the use of savings and debt and believe that actual consumption is a more meaningful indicator.
2. *Relative deprivation approach.* This approach focuses on poverty. If individuals are deprived of the essentials of life then compared to others in society their deprivation level increases and their living standard conversely falls.
3. *Level of living approach.* This approach concentrates on the individual's command over resources as opposed to needs satisfaction. Resource has a broader meaning than merely real disposable income. A high level of living is achieved when individuals have a wide choice of actions upon which to draw to shape their lives and living conditions.
4. *Quality of life approach.* Increases in material affluence are seen as contributing to but not fully explaining improvements in well-being or quality of life. Quantification of the worth of surroundings and lifestyle objectives, or in other words the values that an individual might seek to achieve, at different stages in an individual's life cycle are the basis for measurement of living standards under this approach.

Attempts to measure living standards in the past have been hampered by the introduction of complexities such as different classes of individuals in society, regional effects and the impact of various wealth profiles. Yet it is the assessment of subjective concepts that has proven to be the most illusive and controversial.

Quality of life is considered to be the global term for lifestyle satisfaction, including all the aspirations and desires for the future, and the standard of living is seen as a subset based solely on materialistic criteria. The standard of living can thus be defined to encompass only matters of a monetary nature and measured by the value of goods and services held by individuals. The more goods and services people possess the higher is their standard of living. Thus in terms of quality of life the objective factors are effectively separated from the subjective. While individuals might exhibit different standards of living, it is the rate of change that occurs that is perhaps of the most interest.

By defining the standard of living as a subset of the quality of life, the value of goods and services acquired by individuals for their own use enables changes in the standard of living to be identified. Furthermore, if this trend is compared with the level of real income that individuals earn then a measure of the ability to afford these goods and services is derived.

As the value of acquired goods and services grows in real terms per person so does the resultant standard of living. Yet increases in expenditure without corresponding increases in income might suggest that people are living beyond their means, perhaps through reliance on credit or by running-down their accumulated savings. Thus an affordability gap can be created. Affordability is a function of an individual's 'access' to goods and services.

11.5 Conclusion

Discount rates based on the weighted cost of capital (i.e. real interest rates) reflect the true time value of money and do not distort the relationship between present and future events unfairly. This approach is fundamental to comparisons over time because it accounts for the costs of finance, either real or imputed. Underlying the exchange rate for the time value of money is actually a zero discount rate. This signifies that normally present and future generations are given equal weight.

But base changes in the prosperity of future generations also need to be considered. Where living standards are expected to rise or fall, as may be measured by the ability to afford the same goods and services in different time periods, then the discounting process must recognize this change and compensate future generations accordingly. Intergenerational equity is therefore achieved by the consideration of the value of costs and benefits relative to future generations. The traditional argument that future prosperity is financed by present profitability is a view that must be tempered by the need for current activity to be sustainable.

References and bibliography

Bellinger, W.K. (1991). 'Multigenerational Value: Modifying the Modified Discounting Method'. *Project Appraisal*, vol. 6, pp. 101–08.

Bowers, J.K. (1997). *Sustainability and Environmental Economics: An Alternative Text*. Longman.

Broome, J. (1992). *Counting the Cost of Global Warming*. White Horse Press.

Brown, S.P.A. (1987). 'The Fairness of Discounting: A Majority Rule Approach'. *Public Choice*, vol. 55, pp. 215–26.

Brownlee, H. (1990). *Measuring Living Standards* (Paper No. 1). Australian Institute of Family Studies, Canberra.

Clayton, A.M.H. and Radcliffe, N.J. (1996). *Sustainability: A Systems Approach*. Earthscan Publications.

Cropper, M.L. and Portney, P.R. (1990). 'Discounting and the Evaluation of Lifesaving Programs'. *Journal of Risk and Uncertainty*, vol. 3, pp. 369–79.

Dobbs, I.M. (1982). 'Discounting, Intergenerational Equity and the Almost-Anywhere Dominance Criterion'. *Futures*, vol. 14, pp. 307–12.

Freeman, A.M. III (1977). 'Equity, Efficiency and Discounting: The Reasons for Discounting Intergenerational Effects'. *Futures*, vol. 9, pp. 375–76.

Gilchrist, M. (1986). 'Home Budgeting: Your Lifestyle's Lifebelt'. *Personal Investment*, vol. 4, no. 7, pp. 48–50, 53–54, 56.

Gyourko, J. (1991). 'How Accurate are Quality-of-Life Rankings Across Cities?'. *Business Review*, The Reserve Bank of Philadelphia, March/April, pp. 3–15.

Hutchinson, A. and Hutchinson, F. (1997). *Environmental Business Management: Sustainable Development in the New Millennium*. McGraw-Hill.

Kenny, M. and Meadowcroft, J. (eds.) (1999). *Planning Sustainability*. Routledge.

Kula, E. (1981). 'Future Generations and Discounting Rules in Public Sector Investment Appraisal'. *Environment and Planning A*, vol. 13, pp. 899–910.

Kula, E. (1988). 'Future Generations: The Modified Discounting Method'. *Project Appraisal*, vol. 3, pp. 85–88.

Langston, C.A. (1994). 'The Determination of Equivalent Value in Life-Cost Studies: An Intergenerational Approach'. PhD Dissertation, University of Technology, Sydney.

Manning, I. (1984). 'Measuring the Costs of Living of Australian Families'. *SWRC Reports and Proceedings*, no. 43, April, Social Welfare Research Centre, University of New South Wales, Sydney.

Mishan, E.J. (1988). *Cost-Benefit Analysis* (4th edition). Allen and Unwin.

Nagle, G. and Spencer, K. (1997). *Sustainable Development*. Hodder & Stoughton.

Ng, Y.K. (1983). *Welfare Economics: Introduction and Development of Basic Concepts*. Macmillan.

Nijkamp, P. and Rouwendal, J. (1988). 'Intergenerational Discount Rates in Long-Term Plan Evaluation'. *Public Finance*, vol. 43, pp. 195–211.

Page, T. (1977). 'Discounting and Intergenerational Equity'. *Futures*, vol. 9, pp. 377–81.

Pearce, D.W., Markandya, A. and Barbier, E.B. (1989). *Blueprint for a Green Economy*. Earthscan Publications.

Price, C. (1989). 'Equity, Consistency, Efficiency and New Rules for Discounting'. *Project Appraisal*, vol. 4, pp. 58–64.

Price, C. (1993). *Time, Discounting and Value*. Blackwell.

Tisdell, C. (1993). *Environmental Economics: Policies for Environmental Management and Sustainable Development*. Elger.

Young, M.D. (1993). *For Our Children's Children: Some Practical Implications of Inter-Generational Equity and the Precautionary Principle*. Occasional Publication No. 6, Resource Assessment Commission, Canberra.

Zwartz, S. and Marcus, D. (1989). 'Living Standards'. *Consuming Interest*, no. 41, pp. 4–9.

12

The measurement of sustainability

In normal life people are faced with decisions on a regular basis. The process of decision-making involves identifying, comparing and ranking options using multiple criteria. Often this process occurs without even thinking much about it. However, for large investment-related problems there is a tendency to simplify the decision process into a single monetary criterion. The selected option is the one that demonstrates the highest monetary value.

Project appraisal techniques are employed to structure the diverse array of data into a manageable form and provide an objective and consistent basis of choosing the best solution for a situation. Cost-benefit analysis (CBA) is a respected appraisal technique that is widely used by both public and private organizations to aid in complex investment decisions. In conventional CBA much effort has been made to assess the input costs and output benefits by means of a market approach. With the increasing awareness of possible negative external effects and the importance of distributional issues in economic development, the usefulness of CBA has become increasingly controversial. In response to this dissatisfaction, recent attention has been paid to multi-dimensional evaluation approaches (Nijkamp et al., 1990).

The identification of value for money on development projects is still related to monetary return. But other issues are also relevant, particularly for social infrastructure projects, and some are becoming increasingly significant. For example, issues such as welfare enhancement and resource efficiency are vital to the assessment of sustainable development in the wider social context. Since no single criterion can adequately address all the issues involved in complex decisions of this type, a multi-criteria approach to decision-making offers considerable advantage.

Social costs and benefits (including those related to environment impact) need to be integrated into the evaluation process and a strategy developed that will give these factors proper consideration in practice. Social costs and benefits should not be discounted alongside conventional cash flows as they bear little relationship to financial matters and do not reduce in importance exponentially over time. In fact, future generations may value environmental issues more highly than the present generation.

A multi-criteria approach is better suited to deal with the complexities of sustainability assessment. Key criteria such as investment return, welfare enhancement, energy usage and loss of habitat can be combined to form a single decision model that can rank options according to the perceived level of achievement of sustainable development objectives. Such an approach moves away from conventional economic evaluation and the inherent problems associated with monetary assessment.

12.2 Problems with conventional economic evaluation

Cost-benefit analysis is a technique commonly used to compare, evaluate and rank alternative investment projects that involve cash inflows and outflows over a number of years. There are two established types of cost-benefit analysis: economic and social. Economic analysis involves real cash flows that affect the investor. Social analysis involves real and theoretical cash flows that affect the overall welfare of society.

Social cost-benefit analysis uses the concept of collective utility to measure the effect of an investment project on the community. Externalities are assessed in monetary terms and included in the cash flow forecasts, even though in many cases there is no market valuation available as a source for the estimates. Intangibles may be left out altogether. Nevertheless the technique is useful in that it attempts to take both financial and welfare issues into account so that projects delivering the maximum benefit can be identified.

Because social cost-benefit analysis is directed at maximization of collective utility, it is generally employed as an investment technique by government agencies. Furthermore, the amount of research and estimation that the technique demands means that it is feasible for use mainly on large and complex projects, typically those involving infrastructure such as transport, health care, water irrigation, energy supply and the like.

Discounting is the mechanism employed to adjust future costs and benefits into equivalent terms, and is traditionally based on either a capital productivity or time preference philosophy. The discounting process is extended in social cost-benefit analysis to adjust externalities along with the more conventional cash flows. Utility is therefore discounted. For example, the value of lives saved from the introduction of a new hospital diminishes exponentially as time progresses and hence the value of human life itself is effectively worth less in the future than it is today. This is intuitively unsettling.

A common argument in the literature is that development today creates wealth that enriches future generations, enables scientific and technological knowledge to be enlarged and thereby reduces the future's needs for those resources depleted in the process. For example, Mikesell (1977, p. 24) states that 'if the nation desires to maintain the capitalized value of its exhaustible resources, it can save a proportion of the annual return from their exploitation so ... after the resources are completely exhausted, the earnings from the savings reinvested in reproductive capital should have the same capitalized value as the initial value of the exhaustible resources and would provide an annual return for all time'.

But this is a fallacious view. The concentration on maximization of capital wealth can result in a reduction in environmental wealth to the extent that future generations are significantly worse off. Exponential growth is incompatible with sustainable development and so in the longer term must be curtailed. This may involve a reduction in living standards, or at least a reassessment of what constitutes an acceptable standard. As the latter appears preferable to the former, environmental wealth is likely to be more highly valued than capital wealth by future generations.

Price (1993) describes how in the 1970s and 1980s the capacity of cost-benefit analysis to evaluate environmental amenities in cash terms advanced rapidly. The development brought a whole new range of costs and benefits within the purview of discounting. While both official and independent studies often note that the choice of discount rate is controversial, it is rarely suggested that discounting environmental values might be inappropriate. Whenever environmental effects have been enumerated they have invariably been discounted. Goodin (1982) alternatively advocates that the scope of discounting should be limited to goods and services that are tradeable. Discounting of social considerations such as the value of a human life appears even more indefensible.

More frequently the technique of cost-benefit analysis is being called upon to evaluate environmentally-sensitive projects (Field, 1997). It has, however, come under attack from many conservationists because it can justify investments that cause damage to the environment provided that this damage is outweighed by an increase in capital wealth. The quantification of externalities clearly has a great impact on the outcome. In addition, distributional effects are often overlooked, as the acceptance criterion is related to overall utility improvement and hence does not recognize whether this benefit is evenly shared. For this reason projects that are likely to involve an uneven distribution of benefits should be the subject of a separate study designed to expose such effects.

Social CBA is the primary tool used in environmental economics, yet the technique still has significant problems when applied to the evaluation of environmentally-sensitive projects. The most significant problem concerns the fundamental concept of aggregate efficiency. In its basic form social CBA does not adequately account for environmental issues. A major reason for this is the Kaldor-Hicks principle that underlies CBA theory. It states that any type of cost to society is acceptable as long as a project generates greater benefits. Therefore environmental damage is acceptable if benefits, such as increases in capital wealth, are valued more highly.

Externalities often pose significant problems in their monetary translation, and shadow pricing may become necessary. These costs and benefits can be significant and unrealistic assessment can easily distort results. Where externalities are beyond assessment altogether, and hence become intangibles, important aspects of the project may be neglected. Environmental costs and benefits often are considered intangible, and clearly this fails to adequately address the sustainability question within the boundaries of the technique.

No CBA principle prescribes that part of the benefits should actually be reinvested in measures to avoid or compensate for environmental damage, and therefore substitution between capital and environmental wealth is perfect. Yet from a sustainability point of view, it would be preferable to limit environmental damage in absolute terms once a particular threshold has been reached regardless of the increases in capital wealth that may be anticipated.

The financial assessment and discounting of non-monetary issues that do not reflect tradeable goods and services is reasoned to fall outside the purpose that the CBA technique is designed to serve. The value of many externalities is only tenuously connected to capital productivity or time preference criteria. Furthermore, intangibles are clearly unrelated to the payment or receipt of cash and hence do not involve interest or reinvestment, nor are they affected by myopic human behaviour.

12.3 Multi-criteria decision-making

Traditional cost-benefit analysis uses price as the main tool to evaluate projects based on market transaction. However, over the past decade criticism has stemmed from attempts at putting underlying welfare economic theory into practice. It is often difficult or even impossible to improve social welfare in a society if the natural environment continues to be consumed and depleted. The CBA framework ignores and/or under-estimates environmental assets as there are considerable difficulties in measuring all relevant impacts of a project in monetary units (Abelson, 1996).

Intangibles and externalities have become major components of concern in project development decisions, in particular due to the possibility of undesirable effects on the natural environment. The presence of externalities, risks and spillovers generated by project development preclude a meaningful and adequate use of a market approach methodology. When the analysis turns to such effects as environmental quality and loss of biodiversity, it is rarely possible to find monetary variables that can provide a valid indicator. Although effort has been made to arrive at values for social costs and benefits, in practice it is almost impossible to place anything more sophisticated than arbitrary numerical values on such effects. The requirement for incorporating environmental issues into the project appraisal process is increasing, and so the imputation of market prices becomes more and more questionable.

Equity is another consideration in which market evaluation methods perform poorly. Equity distribution is largely overlooked because it is just the total sum of monetarized effects that is used to determine acceptance. The outcome of development can benefit some groups in the community while others are confronted with negative effects. The methodology assumes that so long as a project generates positive return the groups disadvantaged by the decision can be compensated. However no real compensation necessarily takes place in practice. Thus from a social welfare and justice point of view the equity issue has to be dealt with in another way.

Alternatives have been developed to replace CBA completely with other techniques that do not require social costs and benefits to be monetarized. Cost-effectiveness analysis (CEA) and environmental impact assessment (EIA) are leading solutions in this respect. Others have suggested supplementing CBA with a technique that can measure environmental costs in different ways (Nijkamp et al., 1990; Hanley, 1992; van Pelt, 1993; Abelson, 1996). Multiple criteria decision-making (MCDM) is widely accepted to aid the evaluation process in this respect.

MCDM is a technique designed to value two or more criteria and it is particularly useful for those social impacts that cannot be easily quantified in normal market transactions. MCDM transfers the focus of measuring criteria with prices to applying weights and scores to these impacts to determine a preferred outcome. The total scores that reflect the importance of environmental impacts is used to rank project options. MCDM is a more flexible methodology and can deal with quantitative, qualitative and mixed data, while CBA may only handle data of a quantitative nature. In contrast to CBA, MCDM does not impose any limitation in the number and nature of criteria. Therefore, MCDM is a more realistic methodology in dealing with the growing complexity of development decisions.

12.4 Developing a sustainability index

Methods of market-based evaluation are clearly problematic and insufficient in giving full consideration to social effects. Simply using a non-monetary approach to replace or to supplement the monetary approach in project appraisal is also inadequate. A new solution is required to incorporate the strengths of both approaches and hence capture the key elements incorporated in the sustainable development philosophy. Taylor et al. (1993) describe an index system (known as the *Farmer Sustainability Index*) developed for the agricultural industry, that has proven to successfully reflect the degree of sustainable practice among individual farmers. A similar index system for the built environment will enable the sustainability concept to be used in practice.

Sustainable development is the effective balance between economic progress and environmental conservation, and can be characterized as comprising two essential attributes. One attribute deals with economy-centred performance, measured in terms of delivered value for money. The other deals with society-centred performance, measured in terms of improvements to the quality of life. Value for money is defined as the maximization of wealth and the minimization of resources consumed in the process. Quality of life is defined as the maximization of utility and the minimization of associated impact. All four criteria can be combined together into a single decision criterion known as the sustainability index.

Conventional project appraisal techniques measure net social gain (i.e. net present value [NPV]) to determine if a project should proceed. The sustainability index avoids assigning dominance to monetary criteria, and alternatively measures the relative ranking of projects from a more holistic viewpoint. It expresses sustainability performance in terms of individual criteria, assessed using different units, and supports weighting of economy-centred and society-centred attributes to reflect the specific context of the development. For example, private sector development may ascribe more weighting to value for money considerations, whereas public sector projects may ascribe more weighting to quality of life considerations.

In order to ensure that all criteria are allowed to influence a decision, it is recommended that the proportion of either attribute is not to be less than 25%.

Figure 12.1 illustrates the sustainability concept and derives a formula for the objective measurement of the sustainability index. The higher the index, the more sustainable is the project.

12.5 Sustainability criteria

The sustainability index can therefore be described as comprising four separate criteria as follows:

1. Benefit-cost ratio (ratio:1) is calculated as the discounted project income ($) divided by the discounted life-cost ($) when measured over the economic life of the project. Social and environmental factors are not monetarized, so the ratio is reflective of normal economic CBA. The higher the ratio the better. A ratio of 1 indicates the financial break-even point.
2. Social benefit (value score) is calculated using a weighted matrix scoring method. Social benefits (externalities and intangibles) are identified and weighted, and the ability of each alternative to meet these criteria is evaluated using a fixed scale. Environmental effects are ignored. The higher the score the better the performance of the project.
3. Total energy (Gj/m²) is the sum of the embodied energy used in construction and maintenance activities, plus the operational energy consumed over the project's economic life, expressed per square metre of finished floor area. The lower the total energy content the better.
4. Environmental damage (% risk) is the likelihood of environmental disturbance occurring over the economic life of the project. It can be assessed using a risk analysis strategy or derived from interpretation of an environmental impact statement. The lower the probability of environmental damage the better.

Both benefit-cost ratio and total energy concern the effective deployment of resources in society. This is particularly important today as the supply of natural resources is being progressively depleted. The other two attributes focus on the impact of development on living standards. Social benefit refers to the positive contribution of a project in terms of improving living standards, such as time saving and accident reduction arising over the operational life of a project. Environmental damage focuses on the extent to which a project affects environment quality.

Therefore the sustainability index comprises four criteria where each is expressed in different units that are best suited to their quantitative assessment and combined together to indicate the relative sustainability of competing options. Criteria to be maximized are divided by criteria to be minimized, so the higher the index the more sustainable is a project by comparison to its alternatives. Nevertheless, it is important that each criterion is additionally assessed against set benchmarks to ensure a balanced decision is made.

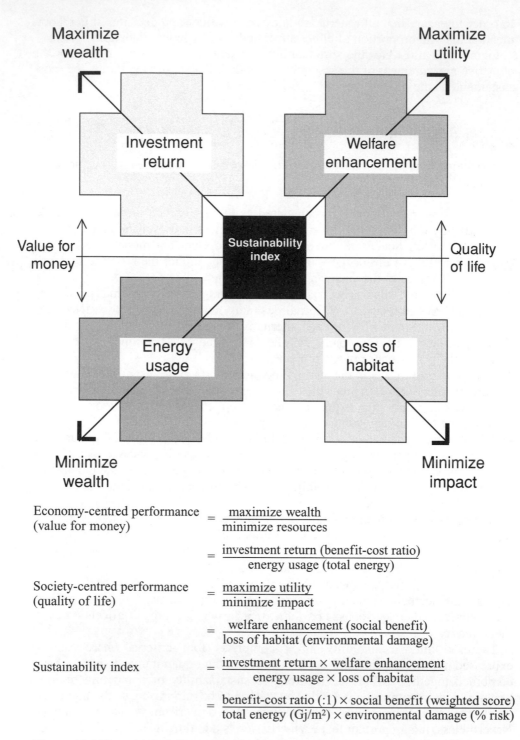

Figure 12.1 The sustainability concept

12.6 Conclusion

Building development involves complex decisions and the increased significance of external effects has further complicated the situation. Society is not just concerned with economic growth and development but is also conscious of the long-term impacts on living standards for both present and future generations. Certainly sustainable development is an important issue in project decisions. The engagement of a conventional single dimensional evaluation technique such as CBA in assisting decision-making is no longer relevant and a much more complicated model needs to be created to handle multi-dimensional arrays of data. The development of a sustainability index is a way to address multiple criteria in relation to project decision-making. The use of a sustainability index will greatly enhance the assessment of external effects generated by property investment and construction, realize sustainable development goals and thereby make a positive contribution to the identification of optimum design solutions.

References and bibliography

Abelson, P.W. (1996). *Project Appraisal and Valuation of the Environment: General Principles and Six Case-Studies in Developing Countries*. Macmillan.

Angelsen, A. (1991). *Cost-Benefit Analysis, Discounting and the Environmental Critique: Overloading of the Discount Rate?* Chr. Michelsen Institute, Department of Social Science and Development, Norway.

Barrow, C.J. (1997). *Environmental and Social Impact Assessment: An Introduction*. Arnold.

Beinat, E. and Nijkamp, P. (eds.) (1998). *Multicriteria Analysis for Land-Use Management*. Kluwer Academic Publishers.

Bell, S. and Morse, S. (1999). *Sustainability Indicators: Measuring the Immeasurable?* Earthscan Publications.

Bowers, J.K. (1997). *Sustainability and Environmental Economics: An Alternative Text*. Longman.

Clayton, A.M.H. and Radcliffe, N.J. (1996). *Sustainability: A Systems Approach*. Earthscan Publications.

Coenen, F.H.J.M., Huitema, D. and O'Toole, L.J. Jr. (eds.) (1998). *Participation and the Quality of Environmental Decision Making*. Kluwer Academic Publishers.

Common, M. (1995). *Sustainability and Policy: Limits to Economics*. Cambridge University Press.

Dragon, A.K. and Jakobsson, K.M. (eds.) (1997). *Sustainability and Global Environmental Policy: New Perspectives*. Edward Elgar Publishing.

Field, B.C. (1997). *Environmental Economics: An Introduction* (2nd edition). McGraw-Hill.

Goodin, R. (1982). 'Discounting Discounting'. *Journal of Public Policy*, vol. 2, pp. 53–72.

Hanley, N. (1992). 'Are There Environmental Limits to Cost Benefit Analysis?' *Environmental and Resource Economics*, vol. 2, pp. 33–59.

Hutchinson, A. and Hutchinson, F. (1997). *Environmental Business Management: Sustainable Development in the New Millennium*. McGraw-Hill.

Kaya, Y. and Yokobori, K. (eds.) (1997). *Environment, Energy and Economy: Strategies for Sustainable Development*. United Nations University Press.

Kenny, M. and Meadowcroft, J. (eds.) (1999). *Planning Sustainability*. Routledge.

Langston, C.A. (1994). 'The Determination of Equivalent Value in Life-Cost Studies: An Intergenerational Approach.' PhD Dissertation, University of Technology, Sydney.

McCornell, R.L. and Abel, D.C. (1999). *Environmental Issues: Measuring, Analysing and Evaluating*. Prentice Hall.

Mikesell, R.F. (1977). *The Rate of Discount for Evaluating Public Projects*. American Enterprise

Institute, Studies in Economic Policy, Washington.

Moldan, B., Billharz, S. and Matravers, R. (eds.) (1997). *Sustainability Indicators: A Report on the Project on Indicators of Sustainable Development*. John Wiley & Sons.

Nagle, G. and Spencer, K. (1997). *Sustainable Development*. Hodder & Stoughton.

Nijkamp, P., Rietveld, P. and Voogd, H. (1990). *Multicriteria Evaluation in Physical Planning*. North-Holland.

Pearce, D.W., Markandya, A. and Barbier, E.B. (1989). *Blueprint for a Green Economy*. Earthscan Publications.

Price, C. (1993). *Time, Discounting and Value*. Blackwell.

Taylor, D.C., Mohamed, Z.A., Shamsudin, M.N., Mohayidin, M.G. and Chiew, E.F.C. (1993). 'Creating a Farmer Sustainability Index: A Malaysia Case Study'. *American Journal of Alternative Agriculture*, vol. 8, no. 4, pp. 175–84.

van Pelt, M.J.F. (1993). *Ecological Sustainability and Project Appraisal*. Averbury.

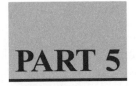

PART 5

Design considerations

Proper design is an important component in the achievement of financial return, social contribution, energy efficiency and minimal environmental impact objectives. It is only through an effective design process that the notions embodied in a feasibility study can be translated into reality. It is therefore crucial that designers are aware of sustainable development concepts and practise them as a matter of routine. Increased development controls, aimed to protect environmental quality, will have the effect of forcing designers to find new ways of delivering client requirements that are neither cost prohibitive nor environmentally harmful.

True sustainability, for the majority of development that takes place, is an unrealistic expectation. However, as designers conceive and implement ideas that approach this goal on a regular basis, society will benefit from new facilities that are less intrusive and which will not become operational burdens for future generations. In time existing facilities that perform poorly will be upgraded to higher standards with lower demand on resource input and greatly reduced waste output.

There are many ways in which built facilities can reduce their dependence on environmental resources. Such strategies range from appropriate material selection to radical designs that embody both passive and active solutions to create more comfortable spaces at reduced cost and energy overhead. Fundamental design considerations like proper siting and orientation, natural lighting and ventilation, insulation and a 'low energy, long life, loose fit' philosophy are now back in fashion and will continue to characterize successful modern development.

A further driving force for sustainable design rests with financial feasibility. In the past, cost advisers have been held responsible for the lack of innovation found in modern facilities because the cost of implementation has exceeded set budget constraints. Now cost advisers are increasingly required to find ways of delivering high quality solutions that satisfy client objectives and simultaneously reduce the total cost of construction and operation. Where operation is inclusive of occupancy costs, designers and cost advisers are realizing that improvements in worker productivity offer potential savings to clients far in excess of those derived from conventional building performance.

Built facilities can become obsolete due to a range of factors, but none are so pervasive as technology. Rapid innovation and change in communication technology has rendered many conventional designs unsuitable, requiring expensive retrofitting. Sustainable practices must include issues like ease of upgrade, flexibility, adaptability and recycling potential, even where these issues are considered to fall to a future owner. In the bigger picture, society is the ultimate loser from poor design.

Design must be understood to comprise the trilogy of form, function and cost. Form reflects the solution adopted to accommodate the intended function and to integrate it with its environment. Function comprises translation of the needs of the client, and in a wider sense the needs of society. Cost includes both initial and recurrent expenditure incurred over the life of the facility. The integration of all three is critical to successful performance and underpins the sustainable development philosophy.

Environmentally-compatible design is sometimes referred to as 'green design' and has led to a perception that it is a fringe movement that at best is an indulgence and at worst an expensive and often ineffective exercise that ultimately has reduced marketplace appeal. But these perceptions are changing as more and more designers are producing highly successful facilities that embody sustainable practices while simultaneously delivering significant financial rewards to their owners. The key to this success is using simple non-mechanical solutions wherever possible and learning from what works well in past designs.

Low energy design is an important part of green design. Heating, ventilation and air conditioning (HVAC) often account for in excess of 40 per cent of construction costs in commercial buildings and represent a high proportion of operating costs in all buildings. The energy required to operate these services, when considered collectively across society, makes a significant contribution to global resource depletion and ecosystem damage. Designs that exhibit reduced dependence on energy for maintaining comfortable internal environments are often linked with high productivity levels and increased satisfaction by occupants. Alternate technologies are providing opportunities for society to use energy more efficiently and hence to prolong natural resources for future generations.

The chapters in this part provide an overview of the impact of built facilities on the natural environment including the use of practical design principles to support sustainable development. Chapter 13 outlines a range of ways in which designers can minimize environmental impact. Chapter 14 investigates passive and active methods for reducing energy dependence. Chapter 15 looks at the important issue of embodied energy within materials and implications for recycling.

Sustainable development involves action at both strategic (macro) and design (micro) levels across the full range of new and existing facilities. It is not something that can be immediately achieved but rather approached slowly. Yet it requires deliberate and persistent action. Professionals in the construction industry are well placed to lead the process of reform. Their ability to do so rests with appropriate education, dissemination of knowledge and a willingness to make a difference.

Environmental impact of buildings

Rick Best

13.1 Introduction

The erection of permanent structures for residential and other purposes is one of the major attributes of civilization. The construction, maintenance and use of these structures all have significant impacts on the environment, both locally and globally. These impacts may produce effects in the short or long term; they may even contribute to irreversible changes in the world's climate, atmosphere and ecosystems. While scientific opinion remains somewhat divided there is an increasing acceptance of the notion that global climate change is happening and that human activity is at least part of the cause (IPCC, 1995). Buildings at all stages of their lives from construction to demolition contribute in many ways to changes, both major and minor, in the environment, many of which are only now being properly identified.

13.2 Visual impact

It is obvious that any building unless completely underground will have some visual impact on the environment. Whether an individual building is judged to be pleasing to the eye or not is largely subjective. It also very difficult to put in place, or implement regulations which will control the appearance of buildings. In many cities it now hardly matters what any single building looks like as the area surrounding it is likely to be so crowded with largely unplanned developments that the designers of new buildings pay little attention to the existing built environment and instead design in isolation.

13.3 Materials

Virtually all building materials have an impact on the environment in some way: from the extraction of minerals such as iron ore and bauxite, to the disposal of demolition materials at the end of a building's life. Much of the impact is related to energy, as energy is used to win and transport raw materials, to process those materials, to fabricate components from the processed materials, to install components, and eventually to disassemble or remove those components, or to demolish the structures of which they were a part. The concept of 'embodied' or 'capital' energy, now the subject of worldwide research, is intended to provide an understanding of the relative impacts of various materials and construction methods in terms of the total energy used in their manufacture and use.

The environmental impact of the processing and manufacture of building materials is not restricted to energy consumption but includes physical degradation around mines, loss of topsoil, loss of forests, destruction of habitat, loss of biodiversity, and depletion of non-renewable resources such as mineral reserves and rainforests. Manufacturing processes produce a variety of toxic and non-toxic wastes, many of which go to landfill or find their way into rivers and groundwater. On-site construction activities produce noise, dust and, sometimes, vibrations which may have significant, although generally short-term, local effects.

Solid waste produced during construction has traditionally been discarded and sent to landfill. This not only adds to the general problems of waste disposal as dumping sites reach capacity and transport distances increase, but also accelerates the rate of exhaustion of non-renewable resources. The disposal of demolition materials has similar impacts. For example, of the 14 million tonnes or so of waste put into landfill in Australia each year, 44% (by weight) is attributed to the construction/demolition industry (McDonald, 1996). The US Green Building Council estimates that each square metre of floor space in the USA generates 12 kg of solid waste from construction alone (Berkebile and McLennan, 1999).

Logging which provides timber for building framing, furniture and finishes is having a devastating effect on forests. In tropical areas the pace of deforestation is alarming – it is suggested that 42 million acres (approximately 17 million hectares) of forest disappear each year (Grant, 1996). Loss of water quality, destruction of habitat, degradation of soils and loss of species (which may contain substances which are potentially beneficial pharmaceuticals) are some of the effects of this forest removal. Forests are also a vital environmental sink and a reduction in forest cover directly reduces the capacity of the Earth to absorb excess carbon dioxide.

Once installed, many materials continue to affect the environment, particularly in respect of indoor air quality (IAQ). Paints, upholstery fabrics, carpets and manufactured wood products are examples of the range of materials that give off potentially harmful gases for a considerable time after installation. Ongoing research is aimed at establishing links between 'outgassing' from building materials and sick building syndrome (SBS). Toxic emissions are not only related to the materials themselves but also to the solvents, finishes and cleaners used to protect and maintain materials after installation.

Many insulating materials used in buildings are made from non-renewable petroleum resources and use CFCs in their manufacture (Bartlett and Prior, 1991). Most of the CFCs used remain in the material and this raises questions of safe disposal or recovery when buildings are refurbished or demolished. Asbestos remains a problem in many existing buildings and requires special procedures for removal and disposal.

CFCs and, more recently, HCFCs have been widely used as refrigerants in chillers and other refrigeration plant. Under the Montreal Protocol the production of CFCs and halons was banned after 1996 but they are still being produced in the developing world and exported to other countries. Most countries are to withdraw 75% of their stocks of CFCs by 2003, while the production of HCFCs will be frozen from 2000 to 2008, with a complete phase-out by 2025. Developing countries were expected to freeze their emissions of CFCs and halons to 1995–97 levels from mid-1999 with a further 50% reduction by 2005 (Planet Ark, 1998).

13.4 Energy

Buildings of all kinds are major energy consumers. The building sector accounts for around one-third of the delivered energy used in most countries, with an even greater portion of electricity use attributable to buildings. Janda and Busch (1994) estimate that 57% of electricity used in developed countries is consumed directly by buildings: 31% in residential buildings, and 26% in commercial buildings.

On-site construction activity requires electricity for tools, lighting, hoists and so on; other items of equipment such as cranes and mixers use fossil fuels that contribute directly to atmospheric pollution. In completed buildings energy is used for many purposes including space heating and cooling, lighting, domestic hot water and to operate various appliances. In Australia, for example, the bulk of this energy is in the form of electricity, mostly generated in large coal-fired thermal power stations. These are generally located close to fuel reserves rather than close to the consumers. Natural gas and fuel oil supply most of the remainder of the energy used with small contributions from renewables such as solar, wind and biomass.

Burning coal to produce steam for electricity generation produces large amounts of greenhouse gases (GHG) as well as causing local environmental damage. Coal is the most carbon-intensive fossil fuel, releasing 29% more carbon per unit of energy than oil and 80% more than natural gas (Planet Ark, 1999). Bush et al. (1995) predict that total Australian carbon dioxide emissions in 2009–10 will be 382 million tonnes of which 183 million tonnes will be attributable to electricity generation. It is also inherently inefficient as only around 40% of the energy contained in the fuel is converted to electricity while the balance is returned to the environment as waste heat. Further losses are incurred when electricity is reticulated from the power station to the consumers.

In 1997 US carbon emissions attributable to the energy sector totalled 532 million tonnes ($\approx$1.95 billion tonnes of CO_2) of which around 86.5% (460 million tonnes) was the result of coal-fired electricity generation (Energy Information Administration, 1998). Nearly 70% of those emissions was actually the result of the production of waste heat, and of the 36% of primary energy use in the US which is devoted to electricity generation, 25% actually only produces waste heat which is dumped into the environment (Spurr, 1996).

13.5 Space conditioning and lighting

The majority of energy used in buildings is devoted to heating, cooling and lighting. In Australia, for example, heating, ventilation and cooling (HVAC) systems account for around 70% of the energy consumed in commercial buildings, and a further 15% is used for lighting (Australian Greenhouse Office, 1999a). Much of the energy is in the form of electricity and therefore contributes to GHG emissions.

Energy use in domestic buildings is divided differently (Australian Greenhouse Office, 1999b) with space heating being the largest single use (38%), followed by electrical appliances, including lights (30%), water heating (27%) and cooking (4%).

HVAC systems in city buildings also contribute to the phenomenon of urban heat islands. This phenomenon occurs when urban areas experience higher temperatures than surrounding rural areas due to factors such as higher heat absorption of paved areas compared with vegetated areas, greater thermal storage of buildings, and the dumping of waste heat from urban areas into the local environment. Building space conditioning systems contribute to this effect by removing heat generated by lights, people and appliances in buildings and pumping that out into the air. The HVAC components themselves generate additional heat, which adds to the amount of heat that is dissipated into the immediate surroundings.

The overall result is that temperatures in urban areas may be as much as 4–5°C higher than those in surrounding rural areas. In extreme cases temperatures may be 6–8°C higher. Electricity utilities' peak demands are increasingly being driven by comfort air conditioning demand during hot summer afternoons. This is exacerbated by the heat island phenomenon.

An increase in global temperatures may actually save energy as demands for heating in colder countries decreases. However a rise of just 2°C in average summer temperature would make a large number of naturally ventilated buildings uncomfortably hot (Bartlett and Prior, 1991) and air conditioning demand would increase.

Lighting in buildings is more or less exclusively provided by electrical means and therefore adds to the greenhouse problem in the same way that any electrical appliance or installation does.

13.6 Stormwater

Runoff from buildings, not only from roofs but also from the paved areas around buildings, is a major environmental problem in urban areas. The quantity of runoff is greatly increased as the patterns of soakage and drainage are disrupted: stormwater which would normally soak into the soil is collected from hard surfaces and then piped away. Natural waterflows are changed, watercourses are filled in, water tables drop, and groundwater supplies downstream are reduced or cut off completely.

Stormwater runoff picks up many pollutants that find their way into rivers and oceans. These pollutants may be solid wastes such as plastics, or chemical substances such as pesticides and fertilizers, which harm fish stocks and aquatic plants or may lead to growths of algae or other unwanted organisms.

Increased runoff can lead to increased soil erosion. More silt is carried into natural waterways with consequent effects on aquatic life; increased deposition in navigable waterways may result in the need for dredging, which in turn can have adverse effects on the local ecosystem.

13.7 Alternative approaches

There has been a dramatic increase in awareness of the environmental problems that we now face. Many researchers and practising design professionals are looking for new ways to satisfy the needs that people have for shelter, comfort and security. These alternative approaches include design for reduced energy consumption, selection of less environmentally damaging materials, waste minimization programmes and the creation of new manufacturing processes for materials which encourage recycling and produce less effluent.

13.8 Natural ventilation and daylighting

As noted earlier, HVAC systems and electric lighting add significantly to the amount of greenhouse gases added to the atmosphere. Naturally ventilated buildings or buildings which use mixed mode systems (i.e. those which use mechanical systems only to supplement or assist natural ventilation when conditions warrant) can reduce that environmental impact yet provide comfortable conditions for occupants. In many cases such buildings are not only less environmentally damaging but also produce economic benefits in the form of improved productivity and substantially reduced running costs (Romm and Browning, 1994).

There are a growing number of examples in the UK and Europe of buildings which demonstrate that air conditioning is by no means essential for the creation of a thermally comfortable indoor environment. Notable examples are The Queen's Building at DeMontfort University in Leicester (Webb, 1994), the Inland Revenue Buildings in Nottingham (Anonymous, 1994), and British Gas' No. 1 Leeds City Office Park, UK (Bunn, 1995).

An outstanding example of the benefits of daylighting is Lockheed's Building 157 located in Sunnyvale, California. The building was designed in the 1980s with extensive use of daylighting rather than artificial lighting. This reduced the energy requirements for lighting by around 75% as well as reducing peak air conditioning demand. Total energy savings were estimated at $500 000 p.a., and the simple payback period at four years. Lockheed believes that savings due to reduced absenteeism which flowed from the creation of a healthier working environment meant that the true payback period was closer to just one year. Productivity gains of up to 15% were also reported (Romm and Browning, 1995).

13.9 Cogeneration

Localizing electricity production and harvesting the heat which is usually considered as a waste from conventional power stations can also provide both economic and environmental benefits particularly if the cogeneration plant is fuelled by natural gas, which produces far less GHG than coal or oil. The efficiency of such power generation facilities is much greater than the large coal-fired stations that are presently used.

The dismantling of energy monopolies and other energy market restructuring in many countries, including the USA and Australia, has started a resurgence of interest in cogeneration (also known as CHP – combined heat and power). When cogeneration based on natural gas is incorporated into an integrated energy management strategy then the environmental impact of associated buildings can be greatly reduced.

13.10 Engineered timber

Engineered wood products including timber I-joists, glue-laminated beams and finger-jointed studs have been available for some time and have gradually gained acceptance (Barreneche, 1994). These products are manufactured from plantation timbers, which are a renewable resource. While they have generally had a higher initial cost than traditional sawn timber these costs could be offset by reduced labour times and reduced material wastage (Anonymous, 1996). More recently, however, the shortage of suitable trees has led to a rapid increase in the cost of large timber sections. On a global scale larger and older trees are becoming increasingly scarce, and to avoid logging these trees to the point of extinction, glued wood products are becoming more popular. The conversion rate for composite products, such as plywoods and hardboards, from tree to useful material may be as high as 90%, compared with only about 40% for conventional solid timber (Rainforest Information Centre, 1998).

Not only do these products reduce the demand for sawn timber and large timber sections, they also allow greater flexibility for designers. Services can be run through the webs of engineered joists, and trusses or large built-up sections can be used which allow greater spans than are generally possible with sawn timber. The lighter weight of structural members made from engineered timber often means that installation can be done by a single tradesman and achieved more quickly.

13.11 Carpets

A lot of carpet is made from nylon or other petroleum-based synthetic materials. They are non-biodegradable and their source materials are non-renewable. Other additives include pesticides, stain treatments, bleaches and dyes. Carpet backing is most commonly a synthetic latex called styrene-butadiene rubber (SBR). The odour, which is typical of new carpet, is caused by emissions of 4-phenylcychexene (4-PC), one of a number of toxic gases emitted by SBR backing.

Wool carpets, although made from a natural, renewable resource, can still have environmental impacts related to overgrazing of land, and the use of toxic dyes and surface treatments. Alternatives are, however, becoming more readily available. Many manufacturers are now producing carpets which give off much less toxic gas (Schomer, 1996) and recycling programmes are gathering momentum with used carpet being diverted from landfill and recycled often as new carpet or some other useful product. In addition a variety of low toxic adhesives and cleaning products are now available.

13.12 Fabrics

New upholstery fabrics are being introduced which set new standards for reduced environmental impact. The McDonough range of fabrics is an excellent example (Anonymous, 1995). These fabrics are the result of a complete analysis and redesign of both the product and the manufacturing process. They are made from a blend of wool from free-range New Zealand sheep, and ramie, a plant fibre similar to linen grown organically in the Philippines. From a list of 4500 dye chemicals used in upholstery manufacture just sixteen were selected that satisfied stringent environmental requirements. With these dyes all colours, except black, can be produced. The fabric is totally biodegradable. Re-engineering of the manufacturing process has significantly reduced waste water production, and even scrap material from the mill is used as mulch by local farmers.

13.13 Waste management

Solid waste is generated both during construction and from demolition, whether as part of refurbishment or complete building removal. Contractors are now becoming aware that by reducing the amount of waste to be dumped a variety of benefits accrue including reduced environmental impact, cleaner and therefore safer sites, and substantial savings on tip fees. In Australia, for example, Fletcher Constructions have achieved notable success with their RECON programme with reductions of up to 50% in the quantity of waste going to landfill and more than 50% reduction in costs associated with waste handling and dumping (McDonald, 1996). By recycling wastes such as plasterboard offcuts, which would previously have been dumped, even greater returns have been achieved.

13.14 Ecolabelling

The extent to which any material or product has an impact on the environment varies at different stages of its life cycle. In a number of countries, notably Germany (and other European Community countries), Japan and Canada, information on the environmental performance of building products is being made available to consumers, designers and specifiers through systems of environmental labelling, or 'ecolabelling' (Atkinson and Butlin, 1993). The EC ecolabel award scheme, for instance, is intended to promote the design and manufacture of products which have a reduced environmental impact over their whole life cycle, and to provide consumers with more information on the environmental effects of products. Only those products that satisfy a stringent set of environmental criteria will be awarded a label.

13.15 Conclusion

Buildings and their multitude of components and systems have serious impacts on the environment. These effects are not limited to the local or immediate environment surrounding a building but include regional and global effects. The environmental impacts of buildings as well as those associated with other human activities (e.g. transport) are assuming greater and greater importance as society accepts the seriousness of the environmental problems that face the world today. The construction industry, which in its widest sense includes designers, demolishers, developers and specifiers as well as contractors, has a duty to address those problems of sustainable development to which it contributes in a great many ways. Material selection, waste management and passive engineering systems are just some of the areas in which the industry can make significant contributions to the general goals of sustainable construction and sustainable development. Some advances have been made but much remains to be done if the environmental impact of the structures which we build are to be reduced to a point where sustainable construction can become an achievable goal.

References and bibliography

Anonymous (1994). 'Energy Efficiency and Commercial Building'. *Constructional Review*, February, pp. 20–25.

Anonymous (1995). 'The William McDonough Fabric Collection'. *Environmental Building News*, vol. 4, no. 8, November/December.

Anonymous (1996). 'The Real Price of Engineered Wood Products'. *The Merchant Magazine*, April.

Atkinson, C.J. and Butlin, R.N. (1993). *Ecolabelling of Building Materials and Building Products*. BRE Information Paper IP 11/93, Building Research Establishment, UK.

Australian Greenhouse Office (1999a). *Australian Commercial Building Sector Greenhouse Gas Emissions 1990–2010*, AGO, Canberra. <http://www.greenhouse.gov.au/energyefficiency/building>

Australian Greenhouse Office (1999b). *Australian Residential Building Sector Greenhouse Gas Emissions 1990–2010*, AGO, Canberra. <http://www.greenhouse.gov.au/energyefficiency/building>

Barreneche, R. (1994). 'Framing Alternatives'. *Architecture*, October.

Bartlett, P.B. and Prior, J.J. (1991). *The Environmental Impact of Buildings*. BRE Information Paper IP 18/91, Building Research Establishment, UK.

Berkebile, R. and McLennan, F. (1999). 'The Living Building'. *The World & I*, vol. 14, no. 10, October, p. 160.

Bunn, R. (1995). 'Natural Speculation'. *Building Services*, vol. 17, no. 11, pp. 20–25.

Bush, S., Holmes, L. and Trieu, L.H. (1995). *Australian Energy Consumption and Production*. ABARE Research Report 95.1, Canberra.

CBEC (1994). 'Energy Efficiency in Commercial Buildings'. Commercial Buildings Energy Code documentation, Melbourne.

Energy Information Administration (1998). *Country Analysis Briefs*. EIA, US Department of Energy. <http://www.eia.doe.gov>

Flavin, C. and Lenssen, N. (1995). *Power Surge*. Earthscan Publications.

Gilchrist, G. (1994). *The Big Switch: Clean Energy for the Twenty-First Century*. Allen & Unwin.

Grant, J. (1996). 'An Opening in the Debate over Forest Use and Conservation'. *Interior Concerns Resource Guide*, ICER, Mill Valley, California. <http://www.numenet.com/intconc>

Herzog, P. (1997). *Energy-Efficient Operation of Commercial Buildings: Redefining the Energy Manager's Job*. McGraw-Hill.

IPCC (1995). Information available at <http://www.usgcrp.gov:80/ipcc/>

Janda, K.B. and Busch, J.F. (1994). 'Worldwide Status of Energy Standards for Buildings'. *Energy*, vol. 19, no. 1, pp. 27–44.

Johnson, S. (ed.) (1993). *Greener Building: Environmental Impact of Property*. Macmillan.

McDonald, B. (1996). 'RECON Waste Minimisation and Environmental Program'. In proceedings of CIB Commission Meetings and Presentations, RMIT, Melbourne, 14–16 February.

Moss, K.J. (1997). *Energy Management and Operating Costs in Buildings*. E. & F.N. Spon.

Planet Ark (1998). 'EU agrees to Phase-out Ozone Destroying Chemicals'. <http://www.planet.ark.com.au/dailynewsstory.cfm?newsid=1893&newsdate=23-Dec-1998>

Planet Ark (1999). 'Environment Group says Coal's Demise has Begun'. <http://www.planet.ark.com.au/dailynewsstory.cfm?newsid=3380&newsdate=27-Aug-1999>

Poole, M. and Prince, R (1995). *A Sectoral Approach to Estimating Gas Turbine Cogeneration Potential in Australian Manufacturing Industry*. University of Sydney.

Rainforest Information Centre (1998). 'Wood-based Composite Building Materials'. *Good Wood Guide*, RIC. <http://www.forests.org/ric/good_wood>

Roaf, S. and Hancock, M. (1992). *Energy Efficient Building: A Design Guide*. Halsted Press.

Romm, J.J. and Browning, W.D. (1994). *Greening the Building and the Bottom Line*. Rocky Mountain Institute, Old Snowmass, Co, USA.

Romm, J.J. and Browning, W.D. (1995). 'Energy Efficient Design'. *The Construction Specifier*, vol. 48, no. 6, pp. 44–51.

Schomer, V. (1996). 'Selecting Environmentally Preferred Carpet and Pad'. *Interior Concerns Resource Guide*, ICER, Mill Valley, California.<http://www.numenet.com/intconc>

Spurr, M. (1996). 'District Energy/Cogeneration Systems in U.S. Climate Change Strategy'. Climate Change Analysis Workshop, Springfield, Virginia, 6–7 June. <http://www.energy.rochester.edu>

Thomas, R. (1999). *Environmental Design: An Introduction for Architects and Engineers* (2nd edition). E. & F.N. Spon.

Tuluca, A. (1997). *Energy Efficient Design and Construction for Commercial Buildings*. McGraw-Hill.

Vale, B. and Vale, R. (1991). *Towards a Green Architecture*. RIBA Publications.

Webb, R. (1994). 'Offices that Breathe Naturally'. *New Scientist*, vol. 142, no. 1929, pp. 38–41.

14

Low energy design

Rima Lauge-Kristensen

14.1 Introduction

Improving energy efficiency is one of the most cost-effective ways of reducing greenhouse gas emissions. Energy use is the dominant source of greenhouse gas emissions, with 55% of total emissions generated by the combustion of fuels to provide stationary energy. Global output of CO_2 is expected to increase by 70% by 2020 compared to 1995 levels and energy levels will rise by 65% in the same period if no mitigating actions are taken. Ninety-five percent of that CO_2 will come from fossil fuel usage (Lardner, 1998).

A building can be designed to use 50% to 70% less energy than a typical building, when the way it functions as a whole and the method in which building systems interact are considered at the concept stage (Sullivan, 1998). The key approach in energy-efficient design is that the building is considered to be more than just cladding and structure, but that all elements contribute to the performance of the building particularly in terms of interacting with the climate. By utilizing natural resources of lighting, heating and cooling, additional energy for these requirements is minimized. Dwellings can be enhanced with energy efficiency principles to allow daylighting, solar winter heating, summer cooling and natural ventilation through proper orientation, thermal storage, insulation, appropriate glazing, cross ventilation or stack effect, external shading and internal zoning.

Operating energy will be further reduced with the use of standard energy-efficient technologies and products (including materials, lamps, ballasts, chillers, motors, variable frequency drives, and glazing). In addition the use of non-toxic materials of low embodied energy drawn from sustainable sources lowers the impact of building materials on the environment. Sustainable design measures are targeted towards reducing energy consumption without compromising comfort, by enhancing occupant health, being resource efficient, cost-effective, and encouraging renewable energy use. Energy-efficient design therefore can satisfy various motivations for alternative building projects, including thermal comfort, cost savings and environmental concerns.

History and vernacular architecture provide many strategies of energy-efficient design and designing for specific climates. For example, the igloo uses local building material configured into the most heat-efficient shape, and the Egyptians and Greeks used solar orientation and thermal mass for natural daylighting and passive heating and cooling (Hsin, 1996). With the invention of artificial lighting, air conditioning and furnaces, many of these practices were abandoned with the result that most buildings consume vast amounts of energy usually derived from nonrenewable resources such as coal, gas and oil and contribute to environmental problems of pollution and global warming.

The awareness of the need to reduce energy consumption and greenhouse gas production from fossil fuel usage is now quite widespread. A great responsibility has been placed on the building and architectural profession as the fruit of their labours can have a large contribution to ecological sustainability. Architects have been gradually embracing ecologically sustainable principles in their designs, a movement formalized in June 1993 at the UIA/AIA Congress of Architects in Chicago, USA. Broadly speaking, this approach encompasses a sensitive respect for the location and users of the building and concerns of conservation of resources and energy (Oppenheim, 1995).

The amount of energy used depends on location, climatic conditions, construction materials, orientation and layout of the building, insulation and number of appliances. A great incentive for the implementation of energy conservation measures is the resulting savings in energy bills. On a greater scale, energy conservation frees up a country's capital for other uses. Utility companies now realize that it is cheaper to improve a customer's energy efficiency and sell the saved energy to new customers than to keep building new facilities in order to increase capacity to provide more electricity and gas. 'It costs up to seven times as much to produce a kilowatt-hour from a new energy source as it does to save a kilowatt-hour through … conservation programs' (Rosenfeld and Hafemeister, 1988).

Many countries now have in place regulations and codes which set minimum requirements in design standards to ensure efficient energy use in new and renovated residential and commercial buildings as a response to the Kyoto treaty agreements to reduce greenhouse gas emissions. The International Council for Local Environmental Initiatives (ICLEI) is 'an association of local governments dedicated to the prevention and solution of local, regional, and global environmental problems through local action' (ICLEI, 2000).

A barrier to implementation of energy-efficient design is the perceived high cost. Many elements of environmentally responsible building do cost more – at least in the short term, but many cost no more than conventional practice (e.g. optimizing solar orientation and accessing prevailing breezes), and in some cases cost less (e.g. reducing window area on east and west façades). In fact when energy savings over time, increased durability, and enhanced worker productivity are factored in, green design features can be most cost-effective (Wilson, 1999).

Figure 14.1 illustrates some of the strategies which can be implemented together to reduce reliance on external sources of energy.

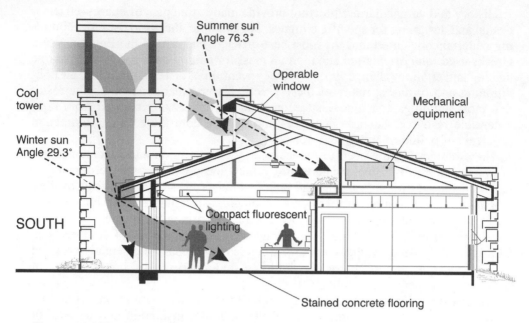

Figure 14.1 The Zion National Park Visitors Centre in Utah
<http://www.caddet-re.org/html/499art4.htm>

14.2 Thermal comfort control

Heat loss or gain through a building shell is dependent on various factors including the type, area and insulating properties of the building element, the temperature gradient imposed on it and wind velocity. The main paths of heat loss in a building are, in decreasing order, the roof, walls, floors, windows, and ventilation/infiltration (Department of Administrative Services, 1989). Many design measures are effective in controlling heat loss and gain for various different climatic conditions. Significant contributions to energy efficiency can be achieved through insulation and passive solar design.

14.2.1 Insulation

An obvious way to regulate the internal temperature of a building is through proper insulation. In cold climates, properly insulated buildings trap the 'free' heat derived from people, lighting, appliances and the sun, thereby delaying extra heating requirements.

Heat flux through walls, floor and ceiling can be reduced by increasing the bulk insulation which influences conductive losses or gains and/or by adding reflective insulation that affects radiant processes. Bulk insulation comes in the form of batts, blankets and boards and is mainly made of glass fibre, wool, cellulose or plastic. The resistance of a given thickness of material or combination of materials to heat flow is given by its R-value – the higher this value, the higher the insulating capacity. Insulating materials are effective by virtue of the trapped air within them which itself has a relatively high R-value. Common reflective insulation materials are reflective foils such as aluminium. They work by reflecting most of the heat on the warm side and not radiating much on the cool side. Reflective insulation is more useful for preventing heat entering rather than escaping a building, and is often used in roofing in warm climates.

Roof insulation in the form of either bulk insulation and/or reflective insulation is most effective when lapped over construction joints between walls and roofs and when fitted tightly between joists and vents. In frame walls, bulk insulation is most commonly used with a vapour barrier included on the warm side to prevent condensation. In cavity brick walls, wire mesh is used to hold the insulation in place so that it does not bridge the gap and create a path for water to transfer to the inside wall. Floor insulation, apart from carpet and underlay, can take the form of bulk insulation fitted between joists. Slab on ground construction has a lower rate of heat loss than timber floors, however because of its high heat capacity it conducts heat away from anything in contact with it and therefore may feel cold to walk on. If this is undesirable, an insulating layer such as cork or carpet may be laid in areas not being used as a thermal mass collector.

Autoclaved aerated concrete (AAC) blocks and similar products provide an energy-efficient alternative to composite brick and timber construction. The trapped pockets of air in the AAC blocks provide them with a good level of insulation. The advantage over other traditional construction systems is that no additional insulation product is required as the cladding is the insulation.

Insulation is not logical in all circumstances; for instance, when walls have excessive windows it may be better to insulate the latter with curtains, tinting, etc. rather than the walls. In hot, humid climates non-insulating walls are preferred to allow quicker cooling during the evening hours.

14.2.2 Passive solar design

The built form of a low energy dwelling is shaped by the climatic conditions of the site. Consideration must also be given to the landscape features of the site and surrounding area and the position of adjacent buildings and trees. In areas or seasons in which large diurnal temperature differences exist or in which heat gain is desirable, various designs are available that will promote the collection, storage and redistribution of heat to maintain even temperatures within the dwelling. In warm climates, strategies in which the heat of the sun is excluded and the cooling effect of breezes is harnessed are more appropriate. Passive solar design principles include orientation of windows relative to the position of the sun, thermal mass, natural ventilation, shading and sunlit areas. Apart from possibly high thermal mass, these features rarely increase the cost of the building as they mainly result from astute planning.

For a building in cold and temperate areas to take maximum benefit of solar access, its long axis should have an east-west orientation and most of the living area windows should face the equator (i.e. north in the case of the southern hemi-

then released into the building through the ceiling at night. An alternative approach is to make a light mass nocturnal radiator, such as a metal roof panel, and move air behind it. The amount of heat that can be rejected overnight is quite modest so night sky radiant cooling is suitable for fairly low heat gain buildings, typically of one storey, in dry climates (Rosenbaum, 1999).

Nocturnal ventilative cooling refers to the strategy of ventilating a building at night, when the ambient temperature is low, and closing it during the day. The structure and the internal air is cooled by the morning which maximizes the potential of the building to absorb heat and delay internal temperature rises during the day. Cool night air can be captured in enclosed external spaces such as courtyards. It is most effective in high mass, well-insulated buildings with low internal heat gains in climates with high daily ambient temperature swings.

Evaporative cooling (effective only in low humidity climates) involves placing a body of water such as a fountain, pool or saturated membrane in the path of a breeze. For water to evaporate, it requires energy that it takes in the form of heat from the air and thus cools it. Planting trees provides shade and a cooling effect caused by the evaporation of the water released by the leaves.

The temperature of the ground three metres or more deep is slightly below the average annual air temperature. Earth-coupled cooling involves partial or full earth-sheltering of a building, or circulating air through earth tubes driven into the ground in either closed or open loops. The basement walls of a building can also be used as an earth-air heat exchanger. Condensation and biological growth may be a problem and proper drainage and treatment needs to be provided (Rosenbaum, 1999).

14.2.6 Natural ventilation

Air movement cools the body as a result of evaporation of perspiration or convective heat losses. In hot humid conditions this is essential for comfort, but in hot arid conditions ventilation may only be desirable in the cooler evening hours. Ventilation cools the interior of a sunlit lightweight uninsulated building but accelerates the rise in temperature in a massive masonry building. In lightweight uninsulated buildings the inducement of air movement to alleviate thermal discomfort is easier than air temperature control. The fixed ventilation of a building should be kept to a minimum so that in extreme conditions it can be shut off to prevent unwanted heat loss or gain.

Air movement in a building may be induced in a number of ways. Buildings can be shaped and orientated to allow for maximum exposure to summer breezes, while the walls, interior layout and window placement are important for efficient circulation of these breezes. Air movement through a space can be induced by thermal buoyancy ventilation (stack effect) whereby warmed air escapes through high openings or vertical airshafts. High and low openings can promote stack effect cooling, but the stack is weak in most places at night. Cross ventilation is the channelling of breezes through openings on opposite external walls and can be enhanced by facing the building at an oblique angle to the prevailing wind. Sizing the inlet area equal to the outlet area and horizontally shaped windows (width greater than height) work best. Protruding wing walls or opened casement windows can act as scoops to enhance wind capture, and can also generate different localized air pressures on the same side of a building, greatly increasing the air flow through the adjacent space (Rosenbaum, 1999).

A naturally forced ventilation system is the solar chimney that creates a draft as the air in it heats and rises and is expelled. The heating efficiency is improved with a black metal covering that absorbs solar heat, and by lining it with thermal mass materials so that it continues to function after sunset. Ventilation occurs even when there is no wind, and is very effective in extracting stale, hot air from within a building. The wind tower, used in many hot arid countries, literally scoops air from the prevailing wind stream. The incoming air is evaporatively cooled as it passes over receptacles of water, and warm air is expelled via leeward openings.

14.2.7 Active heating and cooling systems

Centralized ducted or under floor heating is an efficient system for heating large areas such as a whole house, particularly if fuelled by solar energy or slow combustion stoves. Active solar heating systems, which have either water or air as the circulating medium, use pumps, valves and fans to collect, store and distribute heat, and are themselves powered by solar (or wind) energy. In water systems, energy is stored as hot liquid in a storage tank, whereas in air systems warm air is passed through crushed rocks.

Active evaporative cooling uses the same principle as passive evaporative cooling, but is in an artificial environment with a fan to cause air flow and a water spray or screen. In this form the surface area of the water is increased, which raises the rate of evaporation and therefore the cooling effect. As with heat, coolness can be stored in rock beds. For instance, cold night air is passed through the rocks to cool them down and during the day air is circulated through the rocks and back into the building. Ceiling fans are a good choice in warm humid climates. They raise the comfort envelope several degrees, even at higher humidities, by creating air movement, and the turbulent form of airflow is preferable over uniform air motion.

14.2.8 Windows

As the weakest climatic element of the building envelope, windows are an extremely influential factor in climatic design. Window type, size and placement have tremendous effects on solar heat gains, passive cooling and natural lighting. Glass has a large influence on the internal temperature of a building as it has poor insulating properties, its cold surface can cause draughts (as cold air at the window surface sinks, warm air is pulled down across the glass and cooled) and it offers no shading from the radiation of the sun. Although these characteristics can be used to advantage, they can also be controlled by various means. Well-designed windows and shading devices allow solar heat gain in winter and shade and ventilation in summer while providing enough daylighting. Solar gains achieved by having 60% of the building's windows orientated correctly can reduce the heating load of a house by 25% (Carpenter, 1995). Shutters can be used to control the amount of heat (and light) transferred through the glass, and box pelmets and long wide curtains can limit air movement over the glass and prevent draughts.

In houses, low solar heat gain coefficient (SHGC) glazing should be used on east and west orientations, while high SHGC glazing should be used when passive solar and daylighting are being utilized. In commercial buildings, low SHGC glazing may be appropriate for equator-facing glazing as well, because cooling loads can be high even in the winter, when the sun is low in the sky and heat gain is significant. Large single-pane windows generally offer better energy performance, are more durable, require less maintenance, and cost less than multiple-pane true-divided-light windows. Over a certain size, however, thicker glass requirements will increase costs (Wilson, 1999). The window area should also be appropriately sized to match the available thermal mass in the building.

Super-insulating windows comprising of double or triple glazing have reduced heat loss due to the insulating air space between the panes of glass. The insulating value of a window can be further improved with a thermally non-conductive frame, a thin transparent low emissivity (spectrally selective) coating on the glass, such as tin oxide, which reflects heat back into the building, and by using xenon or argon gas instead of air in the cavity. If a transparent insulator, consisting of a sparse skeleton of glass particles (called Aerogel) is placed in the cavity, the window is given similar insulation properties to a wall (Rosenfeld and Hafemeister, 1988).

Skylights, louvres and clerestory windows can provide many low energy solutions including controlling the heat gain of a building or allowing the escape of hot air during summer.

14.3 Daylighting and energy-efficient artificial lighting

14.3.1 Natural light

Much energy can be saved through provision of natural lighting as an alternative to artificial lighting. Natural daylight is the most desirable source of light as it is not only free but also cooler than most artificial sources. The light-to-energy ratio of sunlight is 100 lumens/watt whereas that of incandescent globes is 15 and fluorescent tubes 50 lumens/watt (Lyons and Lee, 1994). Design features that harness this commodity as well as improving its distribution and control include equator-facing windows, translucent internal walls, mirror-tube skylights and light wells, special collectors and fluorescent guides, and light shelves and 'uplighter' glazing panels to distribute light further into internal spaces by reflecting daylight off the ceiling. Natural lighting can be direct or indirect depending on the desired effect and/or function of the space.

With this light, however, there may be unwanted solar heat gain and glare. This can be addressed with overhangs and low emissivity glazing. Light shelves and laser cut glazing minimize heat gain, yet allow light to penetrate deeply into rooms (Isaacs, 1996). One square metre of clear roof glazing, which provides the same amount of light as 30 fluorescent tubes, allows up to a kilowatt of summer heat – equivalent to a single bar radiator. Compact light-tubes with reflective lining provide much more daylight than conventional skylights of the same size, so they cause fewer heating and cooling problems.

14.3.2 Energy-efficient artificial lighting

In office buildings artificial lighting can account for 45% of the electrical load; in other commercial buildings up to 30%. Of all the energy efficiency measures available, those for lighting are the most affordable and have the shortest payback period (Anonymous, 1997). Due to lower heat emission, more energy-efficient lighting will also reduce cooling requirements particularly in commercial office buildings. The heat created by 2 kW of lighting requires 1 kW of electrical energy in an air conditioning system to remove it.

Fluorescent lights are the most efficient and practical option for most interior applications. Fluorescent lights are at least three times as efficient as low voltage lights and five times as efficient as incandescent lights. Compact fluorescent bulbs use up to 75% less power than incandescent bulbs of the same brightness and last 10 times longer. Fluorescent tubes are available in a wide variety of colours, from a warmer, 'daylight' colour to stark white. Triphosphor fluorescent tubes provide more light per watt and better colour rendering. There have also been improvements in ballasts that are required to start and operate the tubes. Solid state ballasts, which have replaced the standard magnetic ballasts, have been refined to dissipate less power, eliminate flicker and prolong lamp life.

A new generation of the light emitting diode, a bright blue LED, has been developed which can be mixed with existing LEDs to create white light (Anonymous, 2000). A standard incandescent light bulb emits most of its electromagnetic energy in the form of heat and is therefore very inefficient. White LEDs will not only use half as much energy to shine, they will last orders of magnitude longer than the conventional light bulb. Typically, the average life span of an LED is 100 000+ hours (11 years) – that is more than 10–20 times longer than incandescent bulbs. Red LEDs are already being used in tens of thousands of traffic signals all over the world.

Low voltage lighting is not low energy lighting. If used, it should be restricted to critical display applications. Lower wattage globes (for example 20 or 35 watt) are preferable to the more widely used 50-watt low voltage globes.

Light fitting (luminaire) design is also important. For instance, efficient reflectors and prismatic diffusers may increase light output by up to 50% in fluorescent fittings. Opal coverings may reduce it by 25%, and dark tinted covers can reduce it by much more (Department of Industry, Science and Resources, 1995).

It is important to perform a routine maintenance programme to get the maximum performance from a lighting system. For example, dirty luminaires can reduce light output by up to 50%, and old fluorescent tubes, which have run for more than 10 000 hours, may produce 30% less light than new tubes. Fluorescent tubes should be replaced after 70–80% of their rated life to counter lumen depreciation and maintain lighting levels.

14.3.3 Lighting controls

The integration of state-of-the-art lighting controls with energy-efficient lumin-aires can achieve enhanced working or living environments and very high overall savings of up to 85% (Anonymous, 1999a). There is a common misconception that it is more energy efficient to leave fluorescent lights on when they are not needed due to the energy needed to start them and the shortening of lamp life due to the frequent switching. Even if energy conservation measures reduce lamp life by 50% they still produce significant savings – more than enough to make up the differ-ence for additional lamps and maintenance costs (Piper, 1999). Soft-start ballasts reduce normal starting stress and double the number of starts that a lamp can be expected to survive over its operating life, and are particularly beneficial to appli-cations where lamps are started frequently. Studies have shown that implementa-tion of good lighting control practices generally results in a decrease in energy use of 25% to 35% or more (Piper, 1999).

Occupancy sensors save energy and, if well designed, occupants don't even notice that the lights are controlled. Several different technologies of occupancy sensors are available, including direct-line-of-sight devices (passive infrared) and 'see around corner' devices (ultrasonic) and should be suitably chosen to match the situation and coverage area. Lights and equipment linked to key-tag switches can be turned off when the key tag is removed when an occupant leaves the room (Department of Industry, Science and Resources, 1995).

The installation of lighting controls, such as photocells, allow artificial lighting levels to be dimmed if there is sufficient natural light available in the building. Fluorescent tubes, incandescent lamps, low voltage, mercury vapour, metal halide and high pressure sodium lamps fitted with standard ballasts can all be steplessly dimmed, saving up to 45% on electrical lighting expenses and extending the life of lamps (Anonymous, 1999b).

Digital lighting control devices can be installed in a lighting circuit between the meter and the light switches to reduce the energy consumption in a bank of fluor-escent lights by up to 30% with an imperceptible effect on light level and and an increase in lamp-life. Fluorescent lights normally operate at the standard utility supply voltage, but this is only required to allow the lights to strike. The digital device holds the voltage at normal for a preset time and then reduces the voltage by 15%, with only a small reduction in light output. When additional lights are turned on, it switches back to mains voltage until the additional lights warm up. This extends the life of the ballasts and the fluorescent tubes themselves.

14.3.4 Lighting design

Good lighting design in a building, which involves getting the 'right light in the right place at the right time', maximizes the effectiveness of lighting and results in a positive atmosphere (Julian, 1997). Effective lighting is more to do with the way it is distributed and diffused. Dark textured walls can absorb up to 90% of light, while light-coloured walls reflect up to 90% of light. A couple of uplights shining onto white ceilings can illuminate a space better than many lights pointed down onto dark floor coverings. Task lights, along with lower background lighting, can provide plenty of light where it is needed, while saving energy.

Constant light levels can be irritating, particularly when they are inappropriate. People function with an internal biological clock that is run largely by exposure to natural light. As natural light changes it interacts with a person's physiology and affects their general well-being. Lighting systems must be flexible and use all possible natural light.

Although automated energy and light management systems are generally good, a recent study of buildings in the UK found that people were frustrated with the lighting and constantly sabotaged and over-rode the systems to regain some control over their environment (Julian, 1997). As individuals have very different light (and thermal) needs, an effective building management system or lighting system should be flexible and able to be controlled by the users and not by a computer operating on set parameters.

14.4 Appliances

Residential electrical appliances use approximately 15% of energy consumed in the home (Carpenter, 1995). Significant resource conservation would result from using only the most necessary electrical appliances, preferably those with high efficiency ratings. Of the appliances used in the home, the refrigerator is an obvious target for energy conservation as it is in use twenty-four hours a day. Improvements in the design of the refrigerator have led to products that use 50–95% less energy. A more desirable option to lower energy consumption of household tasks is through low energy design features of the house itself. Drying cupboards placed behind a fireplace or a hot water storage tank is an alternative to clothes dryers. Below ground pantries is a traditional way of storing food, especially in cooler climates.

Water heating generally represents a large proportion of energy consumption in the home (up to 50%). Household greenhouse gas emissions could be reduced by up to 70% with a solar water heater compared to an electrical one (Sustainable Energy Development Authority, 1999). Solar water heating is a simple and well-established technology. Water is heated in tubes in flat plate collector panels and is either thermo-siphoned or pumped into a storage tank that is either concealed under the roof or is visible next to the collection plate. To ensure peak effectiveness from a solar hot water system, its orientation and angle of inclination, collection plate area and storage tank capacity should be appropriate to its latitude, climate and usage respectively. Other measures include suitable placement to reduce the number and length of pipes required to the draw off points, insulating the pipes and tank and keeping the thermostat setting as low as possible. Water-conserving fittings, such as showerheads and faucet aerators, not only reduce water use, but also reduce hot water use and energy use.

Photovoltaic systems are at such an advanced level today that they are a viable source of electricity for the home (and office) to operate appliances and lighting. The direct current electricity produced from sunlight can be used to run DC appliances, but is usually converted to AC power. Excess power can be either stored in batteries or sold to the local electricity grid. Photovoltaic systems are either in the form of free-standing solar cell arrays or incorporated into the building casement. Weatherproof solar tiles are available which can be installed instead of normal tiles with no special construction methods required (Editor, 1996), and a photovoltaic array can have a dual function of weatherproof cladding or shading and forms a power plant for a building (Anonymous, 1998a).

High-efficiency heating and cooling equipment are recommended for commercial buildings. Mechanical ventilation is usually required to ensure safe healthy indoor air. Heat recovery ventilators should be considered in cold climates because of energy savings, but simpler less expensive exhaust-only ventilation systems are also adequate. High-efficiency appliances offer both economic and environmental advantages over their conventional counterparts (Anonymous, 1998b).

14.5 Conclusion

Architects and builders have produced many examples of high quality, low energy buildings that cost little more than conventional buildings to construct or refurbish. Owners have reaped the benefits of lower operational costs, greater comfort and better quality indoor environments. Today, many practical means and technologies are available for designing and constructing low energy buildings, including computer software that will predict the thermal and lighting performance of a project before it is even built. The long-term objective is to reduce building energy loads in a cost-effective manner such that renewable sources of energy can meet the energy demands in buildings. This in turn will mitigate the impact of the building and energy sectors on the environment and reduce greenhouse gas emissions.

References and bibliography

Anonymous (1997). 'Lights, Action, Savings'. *Australian Energy News*, Issue 4, June, p. 41. <http://www.isr.gov.au/aen/aen4/4lights.html>

Anonymous (1998a). 'Shadovoltaics'. *Australian Energy News*, Issue 10, December. <http://www.isr.gov.au/aen/aen10/10shado.html>

Anonymous (1998b). 'Checklist for Environmentally Responsible Design and Construction'. *Environmental Building News*, 11 December. <http://www.ebuild.com/Resources/Checklist.html>

Anonymous (1999a). 'Daylight Sensing Dimming Systems Controlling Light'. *Australian Energy News*, Issue 11, March. <http://www.isr.gov.au/aen/aen11/11daylight.html>

Anonymous (1999b). 'Lighting Control Options Energy Conservation Lighting up Savings'. *Australian Energy News*, Issue 13, September. <http://www.isr.gov.au/aen/aen13/13lighting.html>

Anonymous (2000). 'Lighting will Never be the Same'. *Australian Energy News*, Issue 15, March. <http://www.isr.gov.au/aen/aen15/44lighting.html>

Carpenter, S. (1995). 'Learning from Experiences with Advanced Houses of the World'. *CADDET Analyses Series*, No. 14, CADDET, Sittard, The Netherlands.

Department of Administrative Services (1989). 'Energy Efficient Australian Housing'. Commonwealth Government Printer, Canberra.

Department of Industry, Science and Resources (1995). 'Best Practice Ecotourism: A Guide to Energy and Waste Minimisation'. <http://www.isr.gov.au/sport%5Ftourism/publications/bpe/lighting.html>

Editor (1996). 'Out on the Tiles'. *Energy Focus*, vol. 50, p. 30.

Elliott, D. (1997). *Energy, Society and Environment: Technology for a Sustainable Future*. Routledge.

EREN (April 1995). <http://www.eren.doe.gov/erec/factsheets/landscape.html>

Herzog, P. (1997). *Energy-Efficient Operation of Commercial Buildings: Redefining the Energy Manager's Job*. McGraw-Hill.

Hodgson, P.E. (1997). *Energy and Environment*. Bowerdean Publishing.

Hsin, R. (1996). 'Guidelines and Principles for Sustainable Community Design'. Thesis, Florida Agricultural and Mechanical University.
<http://www.state.fl.us/fdi/edesign/news/9607/thesis/thesis.htm#1.3>

ICLEI (2000). The International Council for Local Environmental Initiatives website.
<http://www.iclei.org/>

Isaacs, N. (1996). '1995 Daylight in Buildings Workshop'. *Solar Progress*, vol. 17, no. 1, p. 24.

Julian, W. (1997). 'Illuminated Engineers Design the Right Light'. *Australian Energy News*, Issue 4, June. <http://www.isr.gov.au/aen/aen4/4illuminated.html>

Lardner, P. (1998). 'World Air Pollution to Balloon'. *Planet Ark*, 13 November.
<http://www.planet.ark.org/>

Lyons, P. and Lee, T. (1994). 'Windows, Heat and Daylighting'. *Solar Progress*, vol. 15, no. 2, pp. 6–7.

Mestel, R. (1995). 'White Paint On a Hot Tin Roof'. *New Scientist*, vol. 145, no. 1970, pp. 34–37.

Moss, K.J. (1997). *Energy Management and Operating Costs in Buildings*. E. & F.N. Spon.

Oppenheim, D. (1995). 'Watts in Green Substances: The First ACF Green Home'. *Solar Progress*, vol. 15, no. 4, pp. 40–41.

Piper, J. (1999). 'Lamp Life: Seeing the Light – Does Longer Lamp Life Mean Lower Costs? – That Depends on What you Mean by Lamp Life'. *Building Operating Management*, February.
<http://www.facilitiesnet.com/fn/NS/NS3b9ba.html>

Ristinen, R.A. and Kraushaar, J.J. (1999). *Energy and the Environment*. John Wiley & Sons.

Rosenbaum, M. (1999). 'Passive and Low Energy Cooling Survey'. *Environmental Building News*, October. <http://www.ebuild.com/Resources/Rosenbaum/Low_energy_cooling.html>

Rosenfeld, A. and Hafemeister, D. (1988). 'Energy Efficient Buildings'. *Scientific American*, vol. 258, no. 4, pp. 56–63.

Sullivan, E. (1998). 'The Whole Building Story: A Special Report on Energy Star Buildings and Related Strategies for Improving Energy Efficiency'. *Building Operating Management*, September. <http://www.facilitiesnet.com/fn/NS/NS3b8ij.html>

Sustainable Energy Development Authority (1999). 'At Home'. <http://www.seda.nsw.gov.au/>

Thomas, R. (1999). *Environmental Design: An Introduction for Architects and Engineers* (2nd edition). E. & F.N. Spon.

Tuluca, A. (1997). *Energy Efficient Design and Construction for Commercial Buildings*. McGraw-Hill.

Wilson, A. (1999). 'Building Green on a Budget'. *Environmental Building News*, vol. 8, no. 5, May.
<http://www.ebuild.com/Archives/Features/Low_Cost/Low_Cost.html>

Embodied energy and recycling

Caroline Mackley

15.1 Introduction

Because energy is crucial to every aspect of human activity both now and in the future, issues relating to supply, production and consumption are constantly occupying the headlines of newspapers and debates worldwide. By far the greatest source of energy at present is fossil-based fuel oil, coal and gas. The strong interdependence between energy and human well-being drives a concern that the strictly limited quantities of non-renewable resources will inevitably be exhausted and thus, there is a compulsion to consider substitutions before the existing stocks run out and permanent global climate changes precipitate.

Energy is a lifeline for human activities, but at the same time its waste by-products endanger human life. Waste by-products generated by energy consumption are the main cause of global environmental problems like the enhanced greenhouse effect, urban industrial air pollution and acidification of the environment. The consideration of embodied energy is essential because 50% of the world's annual energy production is consumed by the goods and services supplied to the built environment.

The rationale behind the study of the relationship between embodied energy and goods and services is two-fold. First and most importantly, assessment of the efficiency of a particular operation is only possible through the systematic quantification of the energy flows through a product cycle. Thus, efficiency measures can only be driven once the baseline energy requirements are known. Secondly, the quantification of greenhouse gas emissions (GGE) from processes is only possible on the basis of known energy consumption. In the absence of embodied energy calculations, total carbon dioxide emissions are impossible to calculate.

The significance of the quantification of embodied energy lies primarily in the opportunity to develop and implement mitigation strategies as well as provide a benchmark against which to measure success. Increasing the economic and ecological efficiency of the construction industry will provide exponentially-increasing positive net environmental benefits because of the built environment's all-pervading economic, social and environmental dimensions.

15.2 The essence of energy

Energy analysis is concerned with the assessment of the material conditions in conjunction with these laws in order to minimize the system losses and maximize efficiency. Bullard and Herendeen (1975, p. 268) observed that 'when you consume anything, you are consuming energy'. Historically, economic growth and development are linked to cheap access to abundant fossil energy resources. The dual problem of resource depletion and emissions pollution means that there is a pressing need to drive increasing efficiencies in processes dependent upon non-renewable resources and create incentives to foster growth of alternative low emission energy technologies.

Energy sources are considered to be either primary or secondary, renewable or non-renewable. Primary energy sources are those which can be used in their natural state in order to provide energy, for example coal, oil and natural gas. Secondary energy sources are those that result from the processing of primary sources. Electricity, petroleum and LPG are examples.

Non-renewable resources are those for which it is believed there are finite reserves, and include such fuels as natural gas, oil (including synthetic derivatives), coal and methane (from coal seams), all of which are fossil or carbon based and which liberate carbon dioxide through the combustion process. Renewable fuel sources on the other hand are those without limits, such as water, solar, wind and potentially wave energy. The energy directly produced from these sources is carbon free and without GGE impacts. However, they still have a carbon impact indirectly through the provision of the built infrastructure required to facilitate processing and distribution. Moreover, it is argued that the deforestation required for dam construction results in GGE contribution through reduced carbon sink capacity requiring that the associated GGE be ascribed to the energy produced. Furthermore, there is evidence to suggest that the dam systems utilized by hydro schemes have a finite life due to problems of sedimentation.

Biomass fuels provide a more complex problem. Whilst they are from renewable sources, they are also carbon based, meaning they liberate carbon dioxide and water vapour through combustion. Thus, they are not as 'green' as many would wish them to be.

15.3 Embodied energy

Most commonly embodied energy is defined as 'the energy consumed in all activities necessary to support a process, including upstream processes' (Treloar, 1997, p. 375). Embodied energy is divided into two components, the direct energy requirement and the indirect energy requirement. Direct energy includes the inputs of energy purchased from producers used directly in a process (including, in the case of a building, the energy to construct it). Indirect energy includes the energy embodied in inputs of goods and services to a process, as well as the energy embodied in upstream inputs to those processes (Treloar, 1997, p. 376).

Our interest in embodied energy in relation to the built environment concerns the flows of energy through buildings and their individual components and systems over their life cycle. Strategies aimed at minimizing the energy consumption of buildings has to date been focused on operational considerations, however this focus alone will not affect maximum mitigation of fossil fuel use. Further consideration is necessary of the impact of energy embodied within the specific processes related to the production, manufacture, construction, demolition and disposal of products and services related to buildings.

Traditional thinking dictated that the embodied energy of building construction was relatively small compared with the energy used in its operational life. Contemporary research has now demonstrated that embodied energy can actually comprise up to 50% of the total energy flows of a building over its life cycle (Adalberth, 1994; Pullen, 1996; Treloar, 1996b; Mackley, 1998 and 1999).

The term used to describe the quantification of life cycle energy flows is energy life cycle analysis (ELCA). The ELCA balance for a building includes consideration of the initial capital embodied energy (including all construction materials), the direct site energy for construction, the direct energy consumed in operation and maintenance, the capital and recurring embodied energy for the same maintenance and repairs, and the energy to demolish it and to recycle the salvaged materials. It is internationally accepted that an ELCA must follow the measurement and system boundary methodology established by ISO 14040 and 14042 (International Standards Organization, 1998).

The development of embodied energy intensity extraction methods has significantly advanced the understanding of the importance of embodied energy to the life cycle energy balance of a building and the potential for energy savings through materials selection. Questions remain as to the reliability of embodied energy intensities due to significant variance in published data to date which is largely the result of differences in extraction and reporting conventions. This fact has resulted in the construction industry's hesitance to participate in the application of energy mitigation measures (Mackley, 1999).

15.3.1 Extraction methodologies

There are three main methods of embodied energy intensity extraction: process based, input-output based and hybrid analysis. These methods vary significantly in terms of their system boundaries and reporting conventions, which provides significant confusion for both researchers and the industry. These methods and their relative advantages and disadvantages are briefly discussed below:

1. Process analysis can be described as embodied energy analysis using any other source of information other than input-output tables (Treloar, 1997).

It is generally accepted (Alcorn, 1996; Pullen, 1996; Salomonsson, 1996; Treloar, 1996) that its major advantage relates to the accuracy possible for the precisely defined system to which it relates (e.g. the production of kiln dried timber from a particular mill). However, figures published to date derived by this method suffer from the problem of incompleteness for a number of reasons. These include:

a) Inadequate degree of accounting for direct energy inputs more than two stages upstream from the process due to increasing complexity.

b) The indirect energy associated with the provision of capital goods required for the process both directly and upstream from the process are usually excluded.

c) Direct energy inputs to the process are commonly measured in delivered terms, thus the energy embodied in the delivered energy is excluded from the calculation. 'If primary energy is reported instead of delivered (or end-use) energy, then the value maybe 30% to 40% higher for common building materials' (Pears, 1996, p. 15).

2. 'Input-output analysis utilises a mathematical function known as [the Leontief] matrix inversion, whereby all upstream energy requirements from sectors of the economy are summed' (Treloar, 1997). This provides systemic completeness to the analysis, which is not possible with process analysis. However, problems of validity arise from assumptions that must be made relating to the homogeneity of any or all sectors, the energy tariffs paid by the sectors as well as the prices of the commodities produced from each sector. As most sectors of the economy, as defined and measured by statistics bureaux are not wholly homogenous, significant variations in embodied energy intensities can result when compared to those derived by process analysis from individual manufacturers.

In addition, if the direct energy purchases of each sector are not set to zero, then the resulting intensities suffer from the problem of double counting of inter-energy transactions (Pullen, 1996), where an energy sector contributes to a material input indirectly.

3. Hybrid analysis, on the other hand, 'aims to utilise the completeness of input-output analysis and the reliability of process analysis' (Treloar, 1997). The important embodied energy pathways are extracted using input-output analysis and then specific process data can be used to derive a material intensity without all of the problematic flow-on effects that arise from the two analysis methods individually. The resulting energy intensity is reported in primary energy terms and double counting of direct energy inputs is eliminated. Treloar (1998) claims that this method can provide figures that are up to 90% complete. The problems of validity remain in relation to energy tariffs, primary energy factors and sector homogeneity.

The major advantage offered by this method relates to the consistent quality of data. The resulting energy intensities provide the most complete and constant indicator of the energy required to produce individual materials on a nation-wide basis.

Unfortunately, both pure input-output and hybrid analysis suffer a lack of credibility in the industry due in part to its 'black-box' nature (Treloar, 1996b) and because it provides national industry average material intensities without the efficiency of individual manufacturers being differentiated. This problem ironically is also the method's greatest advantage.

As is evident from the discussion above, it is absolutely essential to the validity of reported embodied energy figures that the method of extraction and the system boundaries are clearly stated. This is particularly so in the case of comparison of embodied energy figures from a variety of sources.

To demonstrate the point in relation to validity, Table 15.1 provides embodied energy intensities for rough processed timber.

Table 15.1 Embodied energy intensity for (rough sawn) timber

Source	GJ/m³	GJ/t	Method	Country of origin
Alcorn (1995)	1.4	2.8	Process	New Zealand
Beca Carter Hollings & Ferner Ltd (1996)	2.7	5.3	Process (DEO)	New Zealand
Forintek (1993)	2.7	5.3	Process (DEO)	Canada
Partridge and Lawson (1995)	1.8	3.6	Process (GER)	Australia
Treloar (1999)	7.0	14.0	Hybrid input-output (D&IE)	Australia
Salomonsson (1996)	6.1	12.2	Process (DEO)	Australia

(Adopted from Table 7 – Salomonsson, 1996, p. 28).

Notes:
DEO   direct energy only accounted for
GER   gross energy requirement but excluding indirect energy
D&IE   direct and indirect energy accounted for
GJ   Gigajoules

Table 15.1 highlights the problem described in the previous discussion. The figures presented vary between the lowest (Alcorn, 1995) of 1.4 GJ/m³ and the highest 7.0 GJ/m³ (Treloar, 1999) by a factor of five. For an average 120 m² brick veneer house, where a total of approximately 8 m³ of timber wall framing is required (excluding timber required for doors and internal trims), this would translate to a variance in the total capital embodied energy figures for timber of between 11.2 GJ in the case of Alcorn, up to 56 GJ for Treloar. Where the average total capital embodied energy for typical residential buildings is between 4 to 6 GJ/m² (Pullen, 1996), the variance provided in the example above would provide an error of between +/– 0.37 GJ/m². In the case of a 4.0 GJ/m² house this would equal approximately a +/– 9.3% error and for a 6.0 GJ/m² house a +/– 6.2% error. The potential for further compounding differences is vast when consideration is given to the multitude of individual parts for any given building.

15.3.2 Additional validity problems

Table 15.1 highlights the variability in published figures to date. It can be seen that the country of origin and date of analysis vary across the data presented. Two further issues arise from this demonstration that further complicate and impact on data quality and validity. These are:

1. *The Primary Energy Factor (PEF)*. This relates to the efficiency with which consumer energy (end-use or secondary) is produced from the primary energy form or fuel. The PEF of a country is applied to the quantity of delivered energy required by a particular process in order to account for all energy inputs in primary terms. The PEF will vary depending on the efficiencies present in an economy's energy production infrastructure and by the nature of the predominant energy production form. For economies such as US and Australia where delivered energy predominates from inefficient coal-fired power stations, the PEF is significantly higher than for countries that have more efficient methods of production (e.g. New Zealand, where hydro-electricity dominates).
2. *Database Date*. As all of these methods rely in part on prices of commodities in the determination of the energy intensity, the year in which a particular calculation is made must be known. For both process and input-output based calculations, a deflator needs to be applied to the intensities in order to adjust them for the current date. In the case of input-output analysis, this is necessary from the outset, as national input-output tables are produced typically only every five years. This is an issue that has been largely ignored by researchers to date.

It is now generally accepted that the most appropriate method of measurement of embodied energy is in primary energy terms. This is seen as an equitable measure, since the impact of the energy conversion process is thus accounted for within the product, process or material in question. This is particularly relevant when consideration is given to the accounting for GGE that result from fossil fuel use.

Most OECD nations have now developed or are developing embodied energy databases. These databases are being designed so that they can be used to quantify the embodied energy of new and old building stocks in an effort to drive further energy efficiencies and mitigate GGEs. The Australian Government, for example, has recognized the importance of this missing link. The 1998 National Greenhouse Strategy (Australian Greenhouse Office, 1998) makes a commitment to pursuing energy life cycle analysis in order to further investigate the link between embodied energy in buildings and the related potential for abatement.

It is important to stress that a systems life cycle approach is desirable where any assessment is made with the objective of improving energy efficiency. In the absence of a life cycle approach, design decisions could be made which result in the deferring of energy flows further up-stream in the life cycle as opposed to eliminating them altogether. Moreover, a systems approach is necessary in order to ensure that the choice of one material over another has a positive quantitative basis.

For example, the Australian extruded aluminium embodied energy intensity is approximately 264 GJ/t with 15.6 tonnes of CO_2 emissions, whereas timber window framing has an energy intensity of approximately 21 GJ/m³ with 2 tonnes of CO_2 emissions (Treloar, 1999). It may not be valid to discount the use of aluminium for all windows in a building on the basis of these figures.

To illustrate this further, if the capital embodied energy (EE) and CO_2 in a fixed window 1200 × 900 mm overall is compared (assuming a 100 × 50 mm frame section, a service life of 50 years and re-painting every 10 years) the figures shown in Table 15.2 would emerge.

Table 15.2 Comparison of timber and aluminium windows

	Timber EE (GJ)	CO_2 (t)	Aluminium EE (GJ)	CO_2 (t)
CAPITAL				
Frame Only	0.93	0.100	2.50	0.40
Paint to Frame	0.10	0.001	0.20	0.02
Subtotal Capital	1.03	0.101	2.70	0.42
OPERATIONAL				
Paint	0.50	0.005	1.00	0.10
Gaskets/Seals	0.96	0.040	0.96	0.04
Subtotal Operation	1.46	0.045	1.96	0.14
ELCA*	2.49	0.146	4.66	0.56

Excluded from consideration above is any theoretical 'credit' that may be achieved through recycling part or the entire frame at the end of its life cycle.

In the case of an average 120 m² three-bedroom residence, where there is 37 m² of windows, the potential energy savings from taking the design decision in favour of timber over aluminium windows is 74.34 GJ and 15.3t CO_2. This example demonstrates the effect of changing the measurement 'system boundary' and the potential downstream effects in the face of ignoring the life cycle view.

15.3.3 Implications for sustainable buildings

The main objective of embodied energy analysis is the development of benchmarks and indicators for energy consumption for buildings using ELCA methodology against which mitigation options can be devised and implemented. The establishment of benchmarks and indicators, consistent with the concept of cost indicators, then provides a benchmark against which mitigation success can be measured.

The logic behind this approach is supported by both traditional cost planning theory and the use of national economic indicators. In relation to cost planning theory, the cost profile of buildings is benchmarked by element and/or trade in order to provide the basis for comparative analysis of like systems, giving due consideration to the cost-benefit of the whole building relative to functional and performance criteria. Moreover, when this methodology is linked with that of life-cost analysis, it provides a broader basis for the quantitative assessment of competing designs, and as such could be considered to be environmental cost planning.

Significant operational end-use efficiencies have been achieved in the building stock of developed nations over the last 10 years by way of legislated minimum thermal performance requirements. However, there is a limit to how far these end-use efficiencies can go. The development of comprehensive embodied energy profiles for buildings will provide the information necessary to enable what is estimated to be a minimum of a 20% reduction in the life cycle embodied energy requirement of buildings (Mackley, 2000).

15.4 Recycling

Recycling is the process of reprocessing materials and treating them as feedstock for the manufacture of new products. The purpose of recycling is two-fold. The first relates to decreased demand for raw materials and energy required in the manufacturing processes and reduction of waste to landfill. The second relates to cost-effectiveness. For example, the purchase of scrap feedstock can be achieved by organizations at a level significantly below the cost of virgin production. Further savings follow through process efficiencies.

From an environmental standpoint, recycling should be promoted on the grounds of increasing the energy efficiency of the reprocessed material. For the major building materials (glass, steel, aluminium and concrete) the majority of the embodied energy is sequestered in the upstream processes such as ore extraction and processing. Thus using recycled material that has already been first and second stage processed, reduces the amount of energy necessary in order to achieve the same mass of virgin product.

Obviously, reducing the energy required for material manufacturing has consequential effects of reducing GGEs, reducing waste streams, minimizing the effects of raw material extraction, etc. Recycling can also deliver other benefits like creating job opportunities or markets in collecting, sorting and processing materials.

Recycling is generally of two types: primary and secondary. Primary recycling refers to the use of recycled materials to make the same or similar products such as aluminium cans, glass bottles, etc. Secondary recycling seeks to use recycled materials to produce new products with less stringent specifications. The down-graded usage of recycled materials (often called 'down cycling') is due to their diminished properties. In terms of value, secondary recycling is considered to be less than that of the primary recycling.

Recycling processes consist of three components: collection, manufacturing and consumption. Collection is a process of recovering materials for reuse from a mixed stream of solid waste. This is the process where recyclable materials are separated from non-recyclable materials. Collection can be either carried out by drop-off programmes or pick-up operations. Drop-off programmes rely on the participation of individuals to deliver materials for recycling to a centralized location. Pick-up operations refer to kerbside collection that are run by volunteer groups or local councils. The recycled materials are collected from the kerb like regular garbage services.

15.4.1 Recycling obstacles

The success of the recycling process relies very much on the participation of individual organizations and industry groups as well as the support of regulatory authorities.

To date, the success of recycled (part or wholly) building materials has been constrained by the 'virgin' paradigm of regulators, owners and industry. There is a perception that recycled also stands for lower quality. Building standards provide a case in point. Australian Standards, for example, provide for virgin materials with no scope for recycled components on the basis of performance measures or criteria.

A good example is concrete. Australian Standards provide only virgin materials for all the components of concrete, both structural and non-structural. Research completed in Australia and New Zealand suggests that the life cycle performance of concrete from recycled materials is similar to the virgin alternative at low cost (BRANZ, 1999; Crentsil et al., 1999; Dumitru et al., 1999). Concrete provides excellent mitigation opportunities as it is one of the top five embodied energy materials (by volume) for buildings. Thus even small measures implemented to increase the recycled component of concrete will have exponential benefits in terms of reduced embodied energy and consequential environmental effects.

Similar problems and solutions exist with glass, aluminium, timber and copper amongst other materials.

The ignorance of industry may also create significant barriers to success. Because typically information pertaining to the environmental effects of building products is ad-hoc, difficult to access and has regional bias, it results in a lack of awareness as well as frustration for industry. Recent research suggests that industry is very interested in the environmental impacts of their activities, however they feel frustrated and discouraged as a result of accessibility problems in relation to unbiased detailed literature (Mackley and MacLennan, 1999).

Complicating this is the inherent conflict of interest of industry in promoting recycling where its main profits arise from established virgin markets. Because of the complex nature of price barriers associated with high energy materials manufacturing, recycling has been kept out of these markets. Where there is a breakdown in these price barriers or price incentives are legislatively facilitated, rapid market expansion results. A good example of this was seen in 1998 in NSW (Australia) when landfill tip fees increased to $50/tonne. The demolition and materials recycling contractors immediately experienced rapid change to minimize the amount of waste taken to landfills. On-site crushing of buildings, re-using the material as road-base and site fill, is an excellent example of the measures being implemented.

Building-related environmental problems will not be solved by recycling alone. Education and information are required to facilitate a paradigm shift away from the conventional throw-away habit or the perception that recycled products are second-hand. Without a combination of these things the market itself is a barrier to recycling. Some product designers (including architects) do not have reuse in mind and do not design their products to be recyclable.

15.5 Conclusion

Strong evidence has indicated that today's environmental problems are closely linked with the consumption of fossil-based fuels. The emission of greenhouse gases during combustion of fossil-based fuels and biomass has become an area of research for energy saving and may lead to a reduction in environmental degradation in the long term. Previously, research has focused on operational energy usage with the embodied energy consumption from initial extraction of raw materials to final delivery being largely ignored. When considering design for energy conservation and saving, all categories of energy for a building over its entire life cycle should be taken into account.

An obvious way of reducing the embodied energy of materials and buildings is though reuse or recycling. Recycling processes are considered to be part of the total solution to environmental problems and can directly contribute to the conservation of limited non-renewable resources. Recycling involves reuse of salvaged materials in manufacture and construction processes. Furthermore, the embodied energy involved in extracting raw materials and delivering them from mining fields to manufacturing plants and ultimately construction sites can be saved as well, and the energy involved in final disposal to landfill can be minimized. As a result, the more materials that are recycled the more energy usage can be minimized for the benefit of future consumption and conservation of the environment.

Since the effects of landfill disposal on the environment have become clear, more attention has been placed in designing for recycling and considering the use of recycled materials for building construction. Apart from eliminating waste disposal problems it also concentrates efforts on energy and pollution reduction. This is not just happening in the construction industry but has also been embraced in the automotive and manufacturing industries by designing for reduced packaging, designing products to last longer, designing products to use either recyclable materials or less virgin materials, and designing products to be fully or partially recyclable.

References and bibliography

Adalberth, K. (1994). 'The Energy Demand from Cradle to Grave for Three Single Family Houses in Sweden'. In proceedings of CIB 1st International World Congress, Watford, England, paper 4, pp. 1–5.

Alcorn, A. (1995). *Embodied Energy Coefficients of Building Materials*. Centre for Building Performance Research, Victoria University of Wellington, Wellington.

Alcorn, A. (1996). 'The Embodied Energy of a Standard House: Then and Now'. In proceedings of the Embodied Energy State of Play Seminar, Deakin University, Geelong, 28–29 November, pp. 133–40.

Australian Greenhouse Office (1998). The National Greenhouse Strategy: Strategic Framework for Advancing Australia's Greenhouse Response. AGO, Canberra.

Beca Carter Hollings & Ferner Ltd (1996). *Forest Industry Energy Research Summary*. Report No. 12, New Zealand Energy Research and Development Committee, Auckland.

Bennetts, J., Oppenheim, D. and Treloar, G. (1995). 'Embodied Energy – Is it Worth Worrying About?' *Solar Progress*, vol. 15, no. 4, pp. 20–21.

BRANZ (1999). *Recycled Concrete Construction Rubble as Aggregate for New Concrete*. Study Report No. 86, Park, S.G.

Bullard, C. and Herendeen, R. (1975). 'The Energy Cost of Goods and Services'. *Energy Policy*, December, pp. 268–78.

Crentsil, K., Brown, T., Bell, D. and Hassan, M. (1999). 'Recycled Concrete Aggregate (RCA): A Survey of Aggregate Quality and Concrete Durability'. In proceedings of Concrete '99 Conference, Concrete Institute of Australia, Sydney, 5–7 May, pp. 277–82.

Curwell, S. (1996). 'Specifying for Greener Building'. *The Architects' Journal*, January pp. 38–40.

Dovers, S. (1994). *Sustainable Energy Systems: Pathways for Australian Energy Reform*. Cambridge University Press.

Dumitru, G., Smorchevsky, G. and Caprar, V. (1999). 'Trends in the Utilisation of Recycled Materials and By-products in the Concrete Industry in Australia'. In proceedings of Concrete '99 Conference, Concrete Institute of Australia, Sydney, 5–7 May, pp. 289–301.

Ecological Sustainable Development Working Groups (1991). *Final Report: Energy Use*. Australian Government Publishing Service, Canberra.

Edwards, P.J. and Stewart, P.J. (1994). 'Embodied Energy Impact Evaluation'. *The Building Economist*, December pp. 21–23.

Edwards, P.J., Stewart, P.J. and Tucker, S.N. (1993). 'Embodied Energy Impact Modelling for Housing Design'. In proceedings of ARCOM 9th Annual Conference, Oxford.

Edwards, P.J., Stewart, P.J. and Tucker, S.N. (1994). 'A CAD-Based Approached to Embodied Energy Impact Modelling for Housing Design'. In proceedings of the National Construction and Management Conference, February, Sydney.

Edwards, P.J., Stewart, P.J., Eilenberg, I.M. and Anton, S. (1994). 'Evaluating Embodied Energy Impacts in Buildings: Some Research Outcomes and Issues'. In proceedings of the First International Conference on Sustainable Construction (CIB TG16: Construction and Waste), November, University of Florida.

Forintek (1993). *Building Materials in the Context of Sustainable Development*. Forintek Canada Corporation.

Howard, N. and Roberts, P. (1995). 'Environmental Comparisons'. *The Architects' Journal*, September.

International Standards Organization (1998). AS/NZS ISO 14040: 1998 Environmental Management Life Cycle Assessment Goal and Scope Definition and Inventory Analysis, ISO.

Lawson, W. (1995). 'Embodied Energy of Building Materials'. In BDP *Environment Design Guide* (PRO2), Royal Australian Institute of Architects.

Mackley, C.J. (1997). 'Life Cycle Energy Analysis of Residential Construction: A Case Study'. Master of Building in Construction Economics thesis, University of Technology, Sydney.

Mackley, C.J. (1998). 'Life Cycle Energy Analysis and its Implications for Sustainable Development'. *The AIQS Refereed Journal*, December.

Mackley, C.J. (1999). 'The Problems with Embodied Energy Extraction Methodology'. Unpublished UNSW coursework for PhD of Built Environment.

Mackley, C.J. (2000). 'Benchmarking Office Buildings for Energy Life Cycle and CO_2 Mitigation Potential'. PhD thesis (in progress), University of New South Wales.

Mackley, C.J. and MacLennan, P.D. (1999). 'Environmental Education for the Building Industry: An Internet Resource'. Research report for CIB W100, November, Massey University at Wellington.

Partridge, H. and Lawson, W. (1995). *The Building Material Ecological Sustainability Index*. Partridge Partners, Sydney.

Pears, A. (1996). 'Practical Policy Issues in Analysis of Embodied Energy and its Applications'. In proceedings of Embodied Energy State of Play Seminar, Deakin University, Geelong, 28–29 November, pp. 15–22.

Powelson, D.R. and Powelson, M.A. (1992). *The Recycler's Manual for Business*. Government and the Environmental Community, Van Nostrand Reinhold.

Pullen, S. and Perkins, A. (1995). 'Energy Use in the Urban Environment and its Greenhouse Gas Implications'. In proceedings of the International Symposium on Energy, Environment and Economics, University of Melbourne, November.

Pullen, S. (1996). 'Data Quality of Embodied Energy Methods'. In proceedings of Embodied Energy State of Play Seminar, Deakin University, Geelong, 28–29 November, pp. 39–50.

Salomonsson, G. (1996). 'Product Comparison Methods'. In proceedings of Embodied Energy State of Play Seminar, Deakin University, Geelong, 28–29 November, pp. 23–31.

Scaife, P. (1994). 'That's Life!'. *Building Product News*, vol. 29, no. 7, p. 14.

Stewart, P.J., Edwards, P.J. and Graham, P. (1995). 'Measuring the Energy Used in the Construction of Residential Housing'. In proceedings of ARCOM 11th Annual Conference, York.

Treloar, G.J. (1995). 'Assessing the Embodied Energy Saving from Recycling and Alternate Materials in Buildings'. In proceedings of SOLAR '95 Conference, Hobart.

Treloar, G.J. (1996a). 'Indirect Embodied Energy Pathways of the Australian "Residential Building" Sector'. In proceedings of CIB Conference, Melbourne.

Treloar, G.J. (1996b). 'Completeness and Accuracy of Embodied Energy Data: A National Model of Residential Buildings'. In proceedings of Embodied Energy State of Play Seminar, Deakin University, Geelong, 28–29 November, pp. 51–60.

Treloar, G.J. (1997). 'An Input-Output Model of the Embodied Energy Pathways of Australian Residential Buildings'. *Journal of Economic Systems Research*, vol. 9, no. 4, pp. 375–91.

Treloar, G.J. (1998). 'A Comprehensive Embodied Energy Analysis Framework'. PhD thesis, Deakin University.

Treloar, G.J. (1999). 'Embodied Energy Intensities of Materials from the Australian Economy for 1992–93 Input-Output Data'. Unpublished database for CSIRO, provided to author by Treloar.

Winter, S. (1991). 'Fuel for Thought'. *Building*, May, pp. 56–57.

PART 6

Energy conservation

Although the environmental performance of built facilities incorporates many issues, energy is an obvious one. The vast majority of energy used in buildings is electricity, commonly generated by nuclear fusion or the burning of fossil fuels. The relative low price of electrical energy is a major reason why building owners and tenants are not overly concerned about further capital spending to reduce energy consumption. The externalities associated with electricity generation, if fully incorporated into pricing models, would significantly raise the unit cost to consumers. The ability of the marketplace to adjust to full recovery of social/environmental costs for electricity is doubtful, politically untenable and all-pervasive in our society.

Conservation of energy through better design solutions is logical but is not a priority with either current or future building owners. This is primarily the result of economic considerations. For energy conservation to be elevated to the top of national policy agenda, it needs to be demonstrated that it can lead to significant cost savings within a relatively short payback period without reduction in the quality of internal environments. This is far from a foregone conclusion.

While operating energy receives some attention because it forms part of the business cash flow and profit determination, embodied energy remains a hidden issue. Yet use of materials which contain high levels of embodied energy is just as socially unacceptable as facilities which are energy expensive to operate. Research into embodied energy is topical, but a large gap remains between current knowledge and its implementation in practice. But if the energy conservation agenda is to be elevated, then designers must understand the impact that their buildings have in energy terms.

The marketplace will demand and seek out energy-efficient solutions if it can be reliably shown that benefits exceed costs. But benefits may be reduced by market apathy and a lack of information available to those making decisions, and costs may be seen to be higher due to the risk of introducing new technology and hidden expenses associated with longer construction periods, reliability and appropriate training. Due to the complications of obtaining perfect information, many designers and building owners make decisions that are sub-optimal. Furthermore, owner-occupiers would judge decisions differently to speculative investors and developers.

The low priority given to energy considerations has implications for the adoption of energy-efficient technologies and processes across the whole construction chain from design to demolition. For example, if commercial tenants are not particularly interested in the environmental performance of their accommodation, such performance will not be a selling point for an investor or developer. Rather, buildings will be designed with features that most future users prize, such as views, comfort, image, acquisition cost and rent level. Location may be uppermost in a consumer's mind, and in this regard the choice of a building to buy or rent will relate to exposure and proximity to services, clients, customers and the like rather than energy efficiency. For this reason energy is not a key design consideration.

It is therefore necessary that some form of government intervention be used to correct inherent market failure caused by investor-centred behaviour. Most countries have moved towards some form of mandatory standards to ensure that new development is not likely to be a resource burden on society in the years ahead. These standards tend to focus on operational energy, but in the future may be more commonly extended to include embodied energy and other sustainability factors. The first step, however, is to ensure that all new buildings granted planning approval meet minimum energy performance standards (i.e. exhibit total energy consumption within legislated maximum limits).

Minimum energy performance standards require research into what level to set benchmarks, and it is important that they are not set so low that they have little effect. Star ratings may be a convenient method of communicating to the public, including prospective tenants, the relative performance of buildings, and may become a key factor in determination of property values and rent levels as consumer awareness of environmental performance increases. Effective education and promotion is vital.

Energy use is bound tightly with the wider issues of sustainability. Whatever approach is used to improve building performance, it is critical that it does not constrain innovation by specifying solutions to achieve required outcomes. Therefore it may be more appropriate to publish maximum total energy limits per m^2 of usable floor space per annum, and leave the method of achieving that goal open to the design team. In this way innovative solutions will be developed to meet the challenges of particular circumstances. Governments may slowly lower maximum limits for new buildings to obtain continual performance improvement. The concept can also be extended to refurbished buildings and existing buildings offered for sale. However, with legislated performance limits comes the issue of enforcement. Perhaps some form of self-assessment linked to the income tax system, together with periodic audits and penalties for excessive energy consumption, are necessary.

The chapters in this part focus on the importance of energy conservation in the built environment. Chapter 16 provides an understanding of the nature of energy and how it is 'consumed'. Chapter 17 looks at renewable energy and technologies that are being used to incorporate it into buildings. Chapter 18 explores contemporary thinking on energy regulation and policy direction applied to built facilities.

Energy is an important part of the sustainable development debate. It measures performance in absolute terms and provides a different perspective to economic modelling techniques. Minimizing energy consumption, whether financially attractive or not, is clearly desirable environmentally. The wider community is the main beneficiary of investment in energy saving features. The aim is to reduce total energy demand, which includes energy embodied in the manufacture of materials, and make buildings more sustainable through lower use of resources and impact on the environment. Issues of enhanced greenhouse effect, global warming and exponential use of non-renewal resources require that the government act to reduce the demand on energy.

Energy quality

Jack Greenland and Rick Best

16.1 Introduction

All deliberate human activity, from taking a train to teasing a tiger, is undertaken on the assumption that the rewards outweigh the risks. It is probably more meaningful simply to rewrite 'rewards' as 'benefits', and 'risks' as 'costs', because in each case the latter is easier to quantify. Of the many ways in which the history of civilization may be written, one of the simplest and most consistent is the account of humankind's use and misuse of energy, which certainly can be described in cost-benefit terms.

As the twenty-first century commences, problems associated with energy and its disposal are rapidly becoming the most discussed technical issues, and are clearly the leading candidates for extensive research and rigorous attention in the coming decades. Unfortunately the debate on energy has, to date, been confused by a simplistic approach to measuring the quantity being 'consumed' (the benefit), and the questionable value (the cost), assigned to it.

It is of use to look at some simple facts about energy, and how these facts are currently being addressed. They are the laws of physics that are not subject to revision in the space and time scale in which we now exist. In other words, they are not negotiable.

16.2 The laws of thermodynamics

Energy cannot be created or destroyed. Its form may be changed, but its magnitude persists. This is the essence of the First Law of Thermodynamics, more commonly known as the Law of Conservation of Energy. No cases of its violation have ever been observed. It follows, therefore, that on receipt of a bill from an electricity distribution authority, charging for 'energy consumed', any self-respecting pedant would return the bill, pointing out that according to the First

Law of Thermodynamics there is just as much energy left after it has been through the system as was supplied at the power point.

As energy cannot be 'consumed' in the strict sense of the term, it follows that it cannot be conserved, or for that matter, wasted.

Something about energy, however, has been changed in the act of processing it by the user. It is a matter of experience that the energy left over after use, although equal in magnitude to that existing before use, cannot be used again for the same purpose. The Second Law of Thermodynamics recounts that a property called 'energy grade' or 'energy quality' has been changed irreversibly. The use of energy is a one-way street. Throughout the history of the Universe, energy, although remaining constant in magnitude, has been degraded, and this process continues.

The simplest measure of energy grade is temperature, and experience shows that the most common forms of heat engines operate by 'taking in' energy at a high temperature (called source temperature) and 'giving it out' at a low temperature (called sink temperature). The larger the gap between source and sink, the greater will be the efficiency of the engine. Indeed a fundamental property of engines, called Carnot Efficiency, determines the maximum possible efficiency of any heat engine.

$$\text{Carnot efficiency } \eta = \frac{T_1 - T_2}{T_1}$$

where: T_1 is source temperature (°K)
                T_2 is sink temperature (°K)

Electrical energy is high grade energy although it possesses no property like temperature. Observe, however, that it can readily be converted into high grade heat.

The inevitable outcomes of these two laws are:

1. Energy cannot be consumed, or recycled. Its quality, however, is continually being degraded.
2. The ultimate fate of all energy processes is the production of waste heat.

These two laws are fundamental properties of the Universe. In the words of Sir Arthur Eddington (1928):

> . . . if your theory is found to be against the laws of thermodynamics, I can give you no hope; there is nothing to do but collapse in deepest humiliation.

16.3 Energy efficiency

We use energy, or more correctly, energy quality, to perform work of many types: to transport people and goods, to heat and cool the places where we live and work, to manufacture and power a multitude of appliances and machines and so on. The three largest users of energy in modern society are transportation, industry and buildings. Demands from heating, cooling and lighting systems in buildings are responsible for roughly one-third of the energy consumed worldwide. Much of that energy is delivered in the form of electricity, with the majority of it being used for heating, ventilation and air conditioning (HVAC) and lighting.

Buildings in developed countries account for 50–60% of electricity use (Janda and Busch, 1994; Berkebile and McLennan, 1999). In Australia 70% of end-use energy consumption in non-residential buildings is devoted to HVAC and 15% to lighting (Australian Greenhouse Office, 1999a); of the energy used for space conditioning 65% is in the form of electricity (CBEC, 1994). In countries where the bulk of electricity is generated by burning coal, this represents a gross waste of energy quality as well as making a significant contribution to greenhouse gas emissions. In Australia, for example, HVAC and lighting account for 84% of commercial building sector greenhouse emissions (Australian Greenhouse Office, 1999a) while energy used in residential buildings for electrical appliances, lighting, cooking and water heating produces over 85% of emissions from that sector (Australian Greenhouse Office, 1999b).

It is now possible to examine some practical examples of benefits and costs in energy terms, in which clear contradictions of interest may be seen.

Take, for example, the task of heating a room for the benefit of the occupants. The costs may be measured in energy terms or in money terms.

Should the room be tightly sealed to prevent infiltration of cold outside air and exfiltration of warm inside air, and should it be thermally insulated against loss of heat by conduction through the walls, efficiency has been improved. It is also possible to reduce the amount of heating required by wearing warmer clothing. It is usual to refer to this kind of attention to thermal efficiency as 'leak plugging and belt tightening'. Here the First Law of Thermodynamics has been addressed, and almost always, energy saving is accompanied by money saving. If the room is heated by electrical resistance heating, however, some other issues become apparent.

At the power station, some 50% of the energy in the coal is lost up the chimney of the generating plant. More energy is lost on the way to the power plug in the room so that usually only one-third of the primary energy is available as delivered energy at this point. Furthermore, in the room, the electrical resistance heater may be replaced by lights, television, refrigerator and other appliances, which ultimately produce the same amount of heat, but perform useful tasks as well.

It can easily be seen that the resistance heater, certainly cheap, is really a frightful waste of energy quality. Here the Second Law of Thermodynamics is being invoked, and money saving cannot always be equated to energy saving. Heating a room using an electrical resistance heater is like using a sledgehammer to drive a panel pin, or using a chain saw to cut butter. Burning coal at temperatures well in excess of 1000°C to produce electricity, which is then used to heat water to less than 100°C is similarly a gross misuse of energy quality.

In summary, the First Law approach requires that a given system be examined, and manipulated or modified to best achieve the required result, in this case that of keeping the occupant at some satisfactory temperature. The Second Law approach requires an assessment of whether the given system is the most appropriate for the task at all. Clearly, both approaches are needed.

Solutions to these issues do not come easily. It is not possible to collect all the heat discharged at the power station, and use it for room heating. It is not really practical to locate televisions, refrigerators, washing machines and the like in a lounge room which needs heating. Yet if a 'business as usual' policy is adopted, the differences between costs in energy terms and costs in money terms will have to be faced, and rationalized.

16.4 Entropy

The concept of entropy was first introduced as a mathematical quantity, part of the attempt by physicists in the nineteenth century to understand the operation of natural irreversible processes. It has come to represent the degree of disorder or randomness of a physical system. Thermodynamic theory states that the entropy of a system can only increase. The entropy of any system can appear to decrease, that is, become more ordered (or less random), but this can only be done by driving the system in reverse. In order for this to happen energy must be applied to the system and when the system is observed together with its surroundings total entropy will be seen to have increased.

To illustrate this notion, consider a piece of coal before it is burned. It has a low entropy value as the energy it contains is high grade and readily available. Once the coal has been burned the level of entropy rises; there is little energy left in the ashes. Energy has not been destroyed: the First Law of Thermodynamics states that this is impossible. The energy originally contained in the coal has, however, become diffuse and is no longer available to do work. The final outcome of this overall increase in entropy, or tendency towards disorder, will be that all the energy in the universe will be converted to waste heat, and temperature will be uniform. While this obviously will not occur for a very long time it is the irreversibility of the process that is important.

The entropy law has also been applied by some economists to theories of resource scarcity and use (Georgescu-Roegen, 1993; Burness et al., 1980; Daly, 1992). The raw materials which are the basis of our industrialized economy are low in entropy, that is, they are in an ordered state and are therefore available to us to use as we see fit. Economic activity and growth depend on the availability of these low entropy resources. As we extract these resources and make use of them, however, entropy increases and we are generally unable to use them again.

Resources with economic value are highly ordered. Iron ore, which consists largely of various oxides of iron, is only useful to us when it occurs in a sufficiently concentrated state such that we can convert it economically into steel. When the steel rusts it returns to the environment as iron oxide but it is now so diffused within the environment that it is no longer economically viable to collect and refine it. Large quantities of energy are converted to heat in the extraction, transport and refining of the ore. Entropy increases both through the production of waste heat from available energy, and through the diffusion of low entropy material resources.

As resources become scarce, market forces, human ingenuity and technological advances may or may not allow economic growth to continue. Some analysts (Cleveland, 1991) believe, however, that the entropy law must constrain economic growth, and that continued economic growth is impossible. Others argue (Peet, 1992) that we must become 'a conserver society', and that labour and capital must be substituted for increased energy use if we are to survive.

16.5 Energy efficiency in buildings

An examination of energy use in buildings, and an analysis of the costs and benefits associated with the adoption of alternative approaches to the problems of providing a comfortable and healthy indoor environment, show that there are a range of benefits which are available with only modest outlays.

The most common approaches to improving energy efficiency in buildings are of the 'leak plugging and belt tightening' type and typically take forms such as window shading, double glazing, weather stripping and the like. These measures can, and do, lead to a reduction in energy consumption and produce money savings but in fact they have nothing to do with energy efficiency, that is, the efficient use of energy. As the Second Law of Thermodynamics demonstrates, true energy efficiency means maximizing the amount of work that is performed as energy quality is degraded.

Application of the Second Law approach requires that primary energy use, i.e. the loss of energy quality, is considered as well as the use of delivered energy measured at the consumer's meter. The importance of this concept as it relates to buildings is demonstrated by studies (Baker, 1992) which have shown that while naturally ventilated office buildings may use only a little over one-third of the delivered energy of fully air-conditioned office buildings, they use less than a fifth of the primary energy. The building owner benefits from lower utility bills (i.e. a reduction in consumption of delivered energy) and society benefits as emissions are reduced and less fuel is consumed.

16.6 Cogeneration

One practical approach to this problem has only emerged as an economic option fairly recently; this is the notion of cogeneration or combined heat and power (CHP). The concept of cogeneration involves localizing electric power generation and capturing or harvesting the waste heat associated with the generation process and employing that heat to do other work.

Conventional coal-fired power stations, such as those commonly used in Australia, are typically located close to their fuel source. As noted above, much of the potential energy in the coal is dissipated as waste heat, and further losses result from the reticulation of the electricity produced from the power station to the consumer. Not only is this inefficient at the source, as the maximum theoretical yield of a thermal power station can never exceed 65.6% (Faucheux and Pillet, 1994), but more energy quality is wasted when the consumer employs high grade energy (i.e. electricity) to perform tasks such as domestic water heating and space heating which require only low grade energy for their satisfactory performance.

In contrast a cogeneration installation, even if using coal as its basic fuel, produces multiple benefits: more work is done by the potential energy of the fuel, and electricity demand is reduced as simple tasks such as heating water are done by direct use of the waste heat from the power generation process. For example, in Denmark (Mortensen, 1995) the efficiency level of coal-fired back pressure and extraction power plants can reach 90% compared with 35–45% for conventional plants. This is achieved by capturing the waste heat from power generation and using it to heat water, which is supplied to houses at 100–115°C for space heating. Waste heat from cogeneration plants can also be used to run absorption chillers, which produce chilled water or ice for cooling.

By utilizing a much greater proportion of the available energy in a fuel, less of the fuel is used and, consequently, harmful emissions are reduced. If natural gas is substituted for coal as the basic fuel, emissions are further reduced as the combustion of natural gas produces significantly smaller amounts of most greenhouse gases.

A typical cogeneration plant may include a gas turbine which is directly linked to a generator, and a steam turbine, which also generates electricity, powered by waste heat captured from the gas turbine (see Figure 16.1). The exhaust from the steam turbine can then be used for domestic hot water or as process steam. In this way the high grade energy in the fuel is degraded in steps with useful work being done at each stage. Other by-products, such as carbon dioxide for industrial use, can be captured as well, providing further benefits.

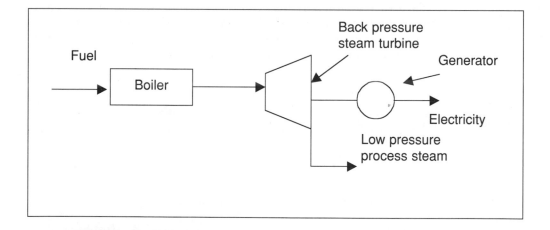

Figure 16.1 A simple cogeneration system producing electricity and steam

16.7 District heating and cooling

A further example of the cogeneration concept which is gaining wider acceptance, particularly in Europe, is the concept of district heating and cooling. This involves central plant facilities producing and reticulating hot or cold water or steam which can then be used for space heating or cooling, or as process heat (see Figure 16.2). In Finland over 40% of the population live in houses which are heated by district heating systems with the majority of public buildings similarly heated. In Helsinki district heating has over 90% of the market (CADDET, 1996).

There are also direct economic benefits which accrue to buildings which are serviced by district heating and/or cooling systems: these include reduced areas of plant rooms (and therefore greater lettable building area), reduced maintenance and running costs of boilers, chillers and other plant as fewer individual units are required, elimination of health risks arising from contamination of cooling towers and other plant no longer required in individual buildings, and improved reliability of service as providers of district services must provide back-up systems in case of breakdowns or during routine shutdowns for maintenance or upgrading of plant.

When cogeneration is combined with careful building design, that is, design which not only takes the 'leak plugging and belt tightening' approach, but which also maximizes the benefits available from measures such as improved daylighting of interiors, utilization of the building fabric for thermal storage, and natural ventilation, then the 'consumption' of energy, in both First Law and Second Law terms, can be drastically reduced.

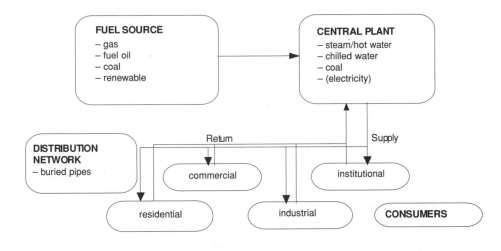

Figure 16.2 Typical district energy system

16.8 Heating and cooling buildings

If using high grade energy sources such as coal and oil to perform 'low grade' tasks is a waste of energy quality, what are the alternatives? One alternative, which has been suggested above, is where heat produced by the combustion of fuel is used successively to perform more than one task. Another is to design buildings in such a way that less energy is required to provide occupants with a comfortable indoor environment.

In some instances integrated systems of ventilation and heating have been adopted where waste heat such as that generated by lift motors, computer installations, and even the human occupants of buildings is gathered and used, via a heat exchange mechanism, to warm incoming fresh air (Holdsworth, 1989). This reduces the energy required to raise the outside air to an appropriate indoor temperature while utilizing waste heat that would otherwise be dumped into the environment. If solar radiation is also gathered the demand for heating energy may be completely eliminated through some parts of the year.

Other strategies which avoid, or minimize, the wasting of energy quality include flushing buildings at night with colder outside air to lower the temperature of the building structure, incorporation of louvres for ventilation, solar gain control and daylighting, and the use of ventilation 'chimneys' which draw heated internal air up and out of the building using natural convection (Webb, 1994; Anonymous, 1996).

These strategies are all aimed at reducing energy demand but, in achieving that reduction, high grade energy sources such as coal are used more slowly, and energy quality is degraded more slowly.

16.9 Conclusion

Energy is essential to all human activity, and is a magnifier of human productivity. Our modern society, which has evolved with great speed since the Industrial Revolution, has become increasingly energy hungry. The bulk of the energy that is used for transport and industry, which are the cornerstones of our society, is converted into usable form by the combustion of non-renewable fossil fuel resources. If our society is to survive then a brake must be put on the consumption of those resources. Energy, by its nature, cannot be conserved or consumed but the rate of consumption of our finite resources can, and must be, carefully controlled.

If 'energy efficiency' is considered only in terms of the First Law of Thermodynamics then only part of the problem of 'energy conservation' is being addressed. A more fundamental approach to energy use is required so that the maximum possible amount of useful work is done by the energy contained in a fuel, and the key to this approach is an understanding of energy quality. Only by examining the grade of the energy available in a fuel, and then using that fuel in the most appropriate way, can we claim to be using energy efficiently.

The most appropriate use can be identified as that which allows the most work to be done with the least loss of energy quality.

References and bibliography

Anonymous (1996). 'Phoenix Fans Flames of Innovation'. *Construction Manager*, vol. 2, no. 5 pp. 7–10.

Australian Greenhouse Office (1999a). *Australian Commercial Building Sector Greenhouse Gas Emissions 1990–2010*. AGO, Canberra. <http://www.greenhouse.gov.au/energyefficiency/building>

Australian Greenhouse Office (1999b). *Australian Residential Building Sector Greenhouse Gas Emissions 1990–2010*. AGO, Canberra. <http://www.greenhouse.gov.au/energyefficiency/building>

Baker, N. (1992). 'Low Energy Strategies for Non-Domestic Buildings'. In *Energy Efficient Building* (S. Roaf and M. Hancock, eds.), Halsted Press.

Berkebile, R. and McLennan, F. (1999). 'The Living Building'. *The World & I*, vol. 14, no. 10, October, p. 160.

Burness, S., Cummings, R., Morris, G. and Paik, I. (1980). 'Thermodynamic and Economic Concepts as Related to Resource-Use Policies'. *Land Economics*, February.

CADDET (1996). 'Helsinki: A Heated District'. CADDET Newsletter – Special Issue on Finland, *CADDET Energy Efficiency*, June pp. 13–14.

CBEC (1994). Energy Efficiency in Commercial Buildings, Commercial Buildings Energy Code documentation, Melbourne.

Cleveland, C.J. (1991). 'Natural Resource Scarcity and Economic Growth Revisited: Economic and Biophysical Perspectives'. In *Ecological Economics: The Science and Management of Sustainability* (R. Costanza, ed.), Columbia University Press.

Daly, H.E. (1992). 'Is the Entropy Law Relevant to the Economics of Natural Resource Scarcity? – Yes, of Course It is!'. *Journal of Environmental Management and Economics*, vol. 23, no. 1, pp. 91–96.

Eddington, A. (1928). *The Nature of the Physical World*. Cambridge University Press.

Elliott, D. (1997). *Energy, Society and Environment: Technology for a Sustainable Future*. Routledge.

Faucheux, S. and Pillet, G. (1994). 'Energy Metrics: On Various Valuation Properties of Energy'. In *Valuing the Environment: Methodological and Measurement Issues* (R. Pethig, ed.), Kluwer Academic Publishers.

Georgescu-Roegen, N. (1993). 'The Entropy Law and the Economic Problem'. In *Valuing the Earth: Economics, Ecology, Ethics* (H.E. Daly and K.N. Townsend, eds.), The MIT Press, 1993, pp. 75–88.

Greenland, J.J. (1993). *Foundations of Architectural Science*. UTS.

Harris, J.M. and Goodwin, N. (1995). *A Survey of Ecological Economics*. Island Press.

Herzog, P. (1997). *Energy-Efficient Operation of Commercial Buildings: Redefining the Energy Manager's Job*. McGraw-Hill.

Hodgson, P.E. (1997). *Energy and Environment*. Bowerdean Publishing.

Holdsworth, B. (1989). 'Organic Services'. *Building Services*, vol. 11, no. 3, pp. 20–30.

Janda, K.B. and Busch, J.F. (1994). 'Worldwide Status of Energy Standards for Buildings'. *Energy*, vol. 19, no. 1, pp. 27–44.

Mortensen, H.C. (1995). 'Combined Heat and Power System in the Greater Copenhagen Area'. *CADDET Energy Efficiency Newsletter*, no. 3, September, pp. 16–18.

Moss, K.J. (1997). *Energy Management and Operating Costs in Buildings*. E. & F.N. Spon.

Peet, J. (1992). 'The Biophysical Systems World View'. In *Energy and the Ecological Economics of Sustainability*, Island Press.

Ristinen, R.A. and Kraushaar, J.J. (1999). *Energy and the Environment*. John Wiley & Sons.

Socolow, R.H. (1975). 'Efficient Use of Energy'. *Physics Today*, vol. 28, no. 8, pp. 23–33.

Tuluca, A. (1997). *Energy Efficient Design and Construction for Commercial Buildings*. McGraw-Hill.

Webb, R. (1994). 'Offices That Breathe Naturally'. *New Scientist*, vol. 142, no. 1929, pp. 38–41.

17

Renewable energy

Rima Lauge-Kristensen

17.1 Introduction

Our society depends heavily on energy for survival and leisure, and the primary source is non-renewable fossil fuels. Since people started farming approximately 12 000 years ago, energy use has been rising and the environmental load we have been placing on the biosphere has increased by more than 10 000 fold (Dovers, 1994). Authorities claim that our energy usage is going to increase four-fold over the next century, and that by 2100 electricity needs will be up to $30–40 \times 10^{12}$ watts worldwide (Integrity Research Institute, 2000). The quality of our future will be largely influenced by our energy use patterns and technological innovations in the power industry. Alternative fuels for electric power generation are receiving attention as a means of reducing harmful emissions (SO_2, NO_x, ash) and greenhouse gases (CO_2, CH_4), decreasing our dependence on limited reserves of fossil fuels and reducing disposal problems.

The greenhouse gas carbon dioxide (CO_2), responsible for 50% of global warming, is predominantly produced by fossil fuel and biomass combustion, particularly in the production of electricity. Approximately 60–80% reduction in fossil fuel use is required to begin to stabilize greenhouse gas concentrations in the atmosphere to safe levels (Integrity Research Institute, 2000). Therefore energy and especially electricity usage must be addressed to attain ecological sustainability. Preferably this should not be at the expense of convenience and comfort, as enforcing reductions in consumption would be unpopular and thus not easily implemented. Rather, technologies to improve efficiencies of energy production and usage, and the use of sources that are renewable and produce minimal wastes should be encouraged.

17.2 Background and history

Up until about 200 years ago, humankind used mainly renewable energy sources. Wood was used for heating, animals for transport and farming, wind for sailing, pumping water and milling grains, and rivers for water wheels – forms of power mostly provided directly or indirectly by the sun. Fossil fuel consumption was generally limited to oil for lighting and coal in foundries and smithies. Examples of more sophisticated utilization of renewable energy include the ancient Greek solar siphon and Roman baths built around hot saline water springs, a practice carried out to this day in various locations. Renaissance alchemists concentrated the sun's energy with concave mirrors for soldering and distillation, and a Swiss scientist, Horace de Saussure, built the first thermal solar collector in 1767, which was later used to heat water and cook food. In 1861, in France, Augustine Mouchot invented a portable solar oven, and solar powered machines which pumped and distilled water and made ice. Solar water heaters, almost identical to modern systems, and solar pumps with parabolic trough concentrators were developed in America in the mid nineteenth century (Rasmussen, 1995). Thousands of years ago the Greeks used water wheels for grinding grain and pumping water, and in the 1800s the water wheel was often used to power machines such as timber-cutting saws in European and American factories. The first hydroelectric plant was established in 1879 at Niagara Falls.

The exploitation of coal started on a large scale in about 1820 (Palz, 1978) with the discovery of huge reserves. The convenience and low cost of fossil fuels stunted the interest in renewable energy sources until the oil shocks of the 1970s when the OPEC countries increased crude oil prices, forcing other countries to resort to other fuels such as coal, nuclear fission and oil shale. New concerns of sustainability and environmental impact have increased support from government and power monopolies to develop renewable energy systems, but this is still hampered because of the entrenchment and low price of electricity produced from conventional systems. However, these differences are diminishing and threats of air pollution, acid rain, oil spills, radiation contamination, global climate change, and considerations about the environment along with its sustainability, are making renewable technology a significant and cost-effective alternative.

Most sources of energy exist as immediate or stored forms of solar energy (except nuclear, tidal and geothermal). Of the fraction of the sun's energy that reaches the surface of the Earth, 30% is reflected as heat, 21% drives the winds, 40% is used in the evaporation-condensation cycle of water and less than 1% is involved in photosynthesis (Dovers, 1994). The sun provides enough energy in one minute to supply the world's energy needs for one year (Alternative Energy Institute, 1999). Therefore direct and indirect solar energy forms have inexhaustible reserves. Significantly, this energy is clean and therefore has little impact on the global ecology (with the exception of hydroelectricity).

The social, economic and technological structure of our society requires vast amounts of cheap and convenient energy, and the task of changing the energy systems of the world is ambitious and formidable. Economic factors will perhaps have the greatest influence on when and how energy alternatives are implemented, as currently fossil fuel is the cheapest and is well established. Although this dependence facilitates modernization and economic development in the present, it could be viewed as short-sighted when sustainability and environmental factors are considered. Depending on the conditions and resources of a region, different forms of renewable energy can be adopted. The majority of renewable energy in the world comes from hydropower and biomass, with advances in photovoltaics (PV) and wind power (Canadian Association for Renewable Energies, 1999).

17.3 Indirect solar energy conversion

Wind energy and hydroelectricity both directly convert the energy of the sun with no chemical, thermal or radioactive waste. Wind and hydroelectric power are the direct result of differential heating of the Earth's surface that leads to air moving about (wind) and precipitation forming as the air is lifted. Photosynthesis products are chemical forms of energy that are usually converted to heat and then into electricity, if desired.

17.3.1 Wind power

Societies have taken advantage of wind power for thousands of years and windmills have been around for almost 1500 years (Alternative Energy Institute, 1999). The Dutch were responsible for many refinements of the windmill, primarily used for pumping water off flooded land. Wind-generated electricity began in Denmark in 1891 (30 kW turbines) progressing to 1980 with the standard features of a concrete tower and three blades with pitchable tips producing 500 kW (Hoagland, 1995). Germany, US, Spain, Denmark and India are among the world's leading nations in the acquisition of wind energy.

High efficiencies (40%) have been made possible by using stronger and lighter materials for the blades to build large output machines (Australian Renewable Energy, 1999). A significant technological development was the variable speed turbine, which rotates at or near the optimum tip speed ratio for any given wind speed providing maximum power extraction. To convert the resulting variable output to a fixed frequency (and voltage), a power converter is fitted between the generator and the grid. A variable speed rotor extracts up to 15% more energy from the wind and makes more use of turbulent winds than a constant speed rotor (McIver et al., 1995). It also reduces material fatigue and maintenance costs as the rotation does not have to be restrained to a fixed frequency. Although large turbines of 1 MW are available, smaller machines continue to be developed, with value engineering making them lighter and more cost competitive. Reduction in the cost of wind power is also achieved by standardized procedures, such as mass production, effective siting and limiting maintenance times during low wind periods (Weinberg and Williams, 1990).

The costs have gone down by more than 80% since the early 1980s so that wind power has become competitive with fossil fuels, and the cost is expected to drop by an additional 20–40% by 2005 (Page and Legerton, 1996). Major offshore developments planned in northern European waters have the potential to drop prices lower than most other forms of renewable energy, as wind speeds are generally higher offshore than on land. Interest in the offshore projects is mainly limited to those countries where there is a shortage of suitable sites on land (Italy and Sweden), or where population density precludes extensive on-land development because of environmental intrusion (Denmark, the Netherlands and the UK).

In countries such as in Australia, Canada, Finland, Japan, New Zealand and Norway, the primary constraint affecting market development is the low cost of conventional generation arising from cheap fuel and surplus capacity. In countries where wind power is economically viable, such as in Denmark, Germany, Italy, the Netherlands, Sweden, the UK and the US, the main constraint is environmental concern (Page and Legerton, 1996).

Various negative concerns about wind farms have been addressed. For example, noise levels have been minimized with improved blade design, and non-metal blades have reduced interference with TV reception. Although bird fatalities are small compared to those caused by oil spills and other fixed structures, the wind industry is increasing safety for birds by such efforts as spacing turbines far apart and in the direction of migration, painting patterns on blades that contrast with landscape colours, and broadcasting a radio frequency to keep birds away. Although wind farms typically take up a large area of land, much of it is farmable. In the UK it has been estimated that the turbines only take 0.01% of the site's land out of agriculture, while roads take a further 1% (Freere, 1995).

17.3.2 Water

The potential energy of water is raised when it is evaporated by the heat of the sun and precipitated above sea level. This potential energy is converted into kinetic or mechanical energy as it flows back to the sea.

17.3.2.1 Running water (river power)

A feasible remote area power supply is the Tyson Turbine that uses the kinetic energy of river flows to rotate a pulley above the water to drive a pump or generator. The water flow rate is increased with a simple system that uses the low head pressure provided by the river flow. A permanent magnet generator, which has low maintenance, produces variable frequency and voltage output that is rectified to DC (direct current) to charge a battery bank. The power output is proportional to the cube of the flow velocity, so that at a water speed of 3 m/s it could supply 2.35 kilowatts of power continuously. Battery storage allows supply during peak demand periods (Singh, 1995).

17.3.2.2 Stored water (hydropower)

Hydropower is not strictly speaking renewable since all reservoirs eventually fill up and require very expensive excavation to become useful again. At this time, most of the available locations for hydroelectric dams are already used in the developed world. Hydropower provides more than 97% of all electricity generated by renewable sources worldwide, and about 20% of the world's total electricity is generated by hydropower. Some regions depend on it more than others. For example, 75% of the electricity produced in New Zealand and over 99% of the electricity produced in Norway come from hydropower.

Hydropower uses water as its 'fuel' supply and the water is released in an unaltered state to continue the natural hydrologic cycle. Hydropower is very convenient because it can respond quickly to fluctuations in demand by opening or closing a dam's gates. Modern hydro turbines can convert as much as 90% of the available energy into electricity, compared to 40% for fossil fuel plants, and in the US, excluding the costs for removing the dam and the silt it traps, hydropower is produced at about one-third of the cost of using fossil fuel or nuclear and one-sixth the cost of using natural gas (Alternative Energy Institute, 1999).

However, hydropower has serious drawbacks, including loss of large areas of land, potential of disastrous dam failures, disruption of river flow and irrigation, altering animal, bird and fish habitats, and methane gas (greenhouse gas) production from rotting vegetation. Long-term environmental hazards are the retention of silt, debris and nutrients. Silt collects behind the dam, accumulating heavy metals and other pollutants that eventually renders the dam inoperable and in need of vast rehabilitation.

17.3.2.3 Ocean power

The harnessing of ocean power is in its infancy, however several projects are in existence. Ocean waves are the result of the transfer of wind energy to the sea. Wave power devices, which can be floating, fixed to the seabed offshore or at a shoreline, absorb this energy and convert it to electricity. Areas with the strongest winds will produce the highest concentrations of wave power. The best areas are on the eastern sides of the oceans (western side of the continents) between the 40 and 60 latitudes in both the northern and southern hemispheres. The waters off California and the UK are regarded as the best potential sites.

Various offshore and shoreline wave power systems exist, including an oscillating water column in which the undulating motion of water compresses air to drive turbines; wave run-up that has a tapered channel system in which waves are captured in a channel and overflow into an elevated lagoon and returned to the sea via a turbine; and sea clams in which air is forced between blades at the perimeter of a circular barge and then through turbines (Wind Energy Training Centre, 2000). Existing projects include the Osprey, a 2 MW shoreline generator developed by Inverness; the Mighty Whale, a floating generator (16–20% efficiency) designed by the Japan Marine Science and Technology Centre in Yokohama which can store energy in the form of compressed air and can also aerate waters, clean up pollution, and calm waters to increase safety (Bickers, 1999); and a system developed in Australia for deep water rock outcrops or harbours which amplifies a wave-front up to three times by focusing it into a parabolic collector which is only 10 metres wide and 8 metres high (Denniss, 2000).

Ocean thermal energy conversion (OTEC) exploits the temperature gradient within the ocean. A temperature difference of at least 20°C from top to bottom, as is found in tropical regions, is required to form the 'heat engine' for a thermodynamic cycle. Warm seawater is evaporated in a partial vacuum or used to vaporize a low boiling point working fluid such as ammonia. This vapour expands through a turbine connected to an electrical generator. The vapour then passes through a condenser that uses cold seawater from the depths of the ocean, forming desalinated water that can be used for other purposes. OTEC taps energy in a consistent fashion and is probably the most environmentally-friendly energy available, but is still largely experimental.

Ocean Power Technology (OPT) of Princeton, New Jersey, has found a way to transform the mechanical force of waves and currents into electricity with piezo-electric polymers. When a piezoelectric material is squeezed or stretched it generates an electric charge proportional to the magnitude of the stress. The hydropiezo-electric (HPE) generator consists of a network of piezoelectric polymer (polyvinylidene fluoride) modules suspended in the ocean between floats and anchors. The polymers are stretched and released as the floats bob in the waves, generating low frequency, high voltage electricity that is sent to shore via submarine cables and converted to alternating current. A 100 MW system covering about 7.7 square kilometres of ocean surface would be capable of powering a city. The polymers are virtually maintenance free, and the modules would be reconditioned after their designed 20-year life span (Geary, 1995).

There are several drawbacks of wave energy, including visual impact and cost. It is also a dangerous obstacle to navigational craft that cannot see or detect it on radar, while fishermen may have trouble with the underwater mooring lines. There are a handful of wave energy demonstration plants operating worldwide, but none produce a significant amount of electricity.

17.3.3 Photosynthesis

17.3.3.1 Biomass

Biomass refers to organic matter that has stored energy through the process of photosynthesis. The most common biomass fuels are wood products, dried vegetation, crop residues and aquatic plants. Biomass has become one of the most commonly used renewable sources of energy in the last two decades, second only to hydropower, probably due to its low cost and indigenous nature. It accounts for almost 15% of the world's total energy supply and as much as 35% in developing countries.

Electricity production from burning biomass to generate steam for turbines is competitive in areas that have agricultural or industrial wastes such as wood chips and sugar cane residues. Gas turbines are generally cheaper and simpler than steam turbines and can be used with biomass to first gasify it with air and steam at high pressures and then burn the gaseous fuel. The hot combustion products are used in a generator to create electricity, while the hot turbine exhaust gases can be used for industrial applications or for additional power generation. Biomass is easier to gasify than coal and contains much less sulphur and ash (Weinberg and Williams, 1990). Fuelling gas turbines with sugar cane residues is a great economic opportunity for electricity co-production in sugar factories and alcohol distilleries. In Varnamo, Sweden, 82–83% efficiency is achieved in a cogeneration plant which gasifies wood to fuel a jet engine to produce 6 and 9 MW of clean power and heat respectively (Yan et al., 1997).

The production and use of biomass has some disadvantages that prevent it from becoming an ideal alternative fuel. For instance, the replacement of coal by biomass could result in deforestation and biomass can only qualify as renewable if it is grown sustainably. In addition it can only be regarded as not contributing to global warming if the carbon dioxide released when burning biomass is balanced by the amount of carbon dioxide used during photosynthesis, i.e. through replacement of the source.

Apart from wood, agricultural residues and energy crops, municipal solid waste, refuse-derived fuel, sewage sludge, paper sludge, manure and solid-waste landfill gas (LFG) is being converted to electricity. For example, municipal waste sites are covered with a geomembrane to trap the LFG resulting from anaerobic digestion (fermentation) of the wet waste. The LFG containing 50–55% methane is extracted to power boiler plants (CADDET, 1997). This alternative form of energy has the additional benefit of processing waste, and many plants are in operation around the world (Venendaal and van Haren, 2000).

17.3.3.2 Fuel production

In Brazil research is being carried out to grow biomass plantations on degraded land and to use the wood or other agricultural wastes to produce fuels such as methanol by gasification or ethanol by fermentation (Hoagland, 1995). A novel renewable fuel with zero greenhouse impact is coconut oil. It has been shown by a collaborative effort of the Australian National University, AcrossTech, CSIRO and University of Wollongong that it can be used to run a unmodified diesel engine with a power output and economy similar to running it on diesel. Transport costs would preclude the oil from becoming available on a widespread level, however it is ideal for remote areas and islands where it is locally grown (Etherington, 1995).

A very different way of converting solar energy into electricity is a system developed at the University of West of England at Bristol in which algae is grown in sunlight, dried and burnt in an engine. The carbon dioxide generated during burning is recycled into the growing algae (Coghlan, 1993).

17.4 Direct solar energy conversion

Solar power is used in two forms, thermal and photovoltaic. The first collects heat from sunlight and is used to warm buildings, heat water, generate electricity, dry crops or destroy dangerous waste. The second produces electricity directly when sunlight strikes a semiconductor material.

17.4.1 Solar heat applications

Approximately three-quarters of domestic and industrial energy use is thermal (Dovers, 1994) and much of this can be sourced directly from the sun. There are generally two types of solar collectors: flat plate collectors for low temperature applications and trough and dish concentrators for greater than 100°C.

17.4.1.1 Solar space and water heating

Solar hot water is a simple and well-established technology. Flat plate collectors are efficient below 70°C as convective and radiative losses are low at these temperatures, which makes them ideal for domestic hot water. Water is heated in tubes in the collector panel and is either thermosiphoned or pumped into the storage tank.

A low-cost solar space heating system would reduce residential fossil fuel use by up to 75% (Dovers, 1994). The shape and tilt of the rooftop collector is optimized for winter and the energy is stored in the form of hot water in a tank. This provides space heating as well as hot water and it can also provide heat for stove-top cooking.

17.4.1.2 Fuel generation

Researchers at the Australian National University have developed a cheap, safe and easy way of storing the energy of the sun in a chemical bond. Energy is focused with a parabolic dish into a chemical reactor containing high-pressure ammonia gas and a metal catalyst, and when temperatures reach approximately 600°C the ammonia is split into hydrogen and nitrogen. Just 400 litres of this mixture, filling a domestic hot water tank, could meet the daily electricity needs of an average family home. The energy of the gas mixture is released by supplying a small amount of heat to start a self-sustaining reaction that produces heat to make steam to drive a generator. An added bonus is that ammonia is one of the most commonly synthesized chemicals and the reactions involved are well understood. Other chemicals can also be used including methane, hydrogen sulphide and metal oxides (Luzzi, 1996). The electricity is not as cheap as conventional power, yet the storage facility makes the technology invaluable particularly in remote areas that rely on expensive diesel-powered generators (Gilchrist, 1995).

17.4.2 Solar electricity

The sun's radiation can be converted into electrical power directly by the use of photovoltaic materials or indirectly by converting the collected heat via a thermodynamic process. Direct conversion systems operate at ambient conditions and use direct and diffused radiation, and therefore can be used in most climates. Thermodynamic conversions occur above ambient temperatures, and concentrators collect only direct radiation and hence are only suited for warm sunny regions.

17.4.2.1 Thermodynamic conversion

Any solar heat collection device along with any available thermodynamic conversion process, such as the highly developed Rankine cycle or gas turbine technologies, can achieve the indirect conversion of solar energy into electricity through heat. Solar collectors track the sun and concentrate the sunlight to heat a fluid that is used in a electricity generating cycle. High temperature solar thermal power plants are most suited for areas with greater than 5 $kWh/m^2/day$. Several different focusing collector technologies exist, but they all make use of reflectors.

The Australian National University has developed a system that uses a paraboloidal mirror dish to focus sunlight on a receiver in which water is converted to steam that is piped to a generator. To maximize output, the tilt and rotation of the dish is computer-controlled to track the sun. This system can be scaled up to hundreds of megawatts (Kaneff, 1994). The large commercial generators in the

Mojave Desert, California, installed by LUZ International, use parabolic trough shaped mirrors that track the sun by rotating about their longitudinal axes. The sunlight is focused onto pipes containing oil that is used to heat steam that drives a turbogenerator (Anonymous, 1998c). Another design is a large field of collectors focused on a central receiver mounted on a tower. Very high temperatures can be achieved for use either in gas turbine heat engines or molten salt storage where heat can be stored for up to six hours. However, this design is not as good as the other systems in that reflector area is used less efficiently and each mirror has to have its own tracking programme. It is also not modular so that scaling up or down would require redesigning all components (Dovers, 1994).

A simple approach is the solar pond, which contains saline water heated by sunlight. The salt increases the density of water so that it sinks to the bottom of the pond, from which it is removed to form steam that drives a Rankine-cycle engine. Fresh water is a by-product and the cool water in the top layer can be used for air conditioning, both useful in hot arid climates (Hoagland, 1995).

A significant advantage of solar thermal energy is that it can use conventional steam power generation equipment, which means that fossil fuel power stations can be converted. Other advantages are that it is suitable for large-scale grid supply and that heat is easier to store (e.g. in thermochemical reactions) than electricity and can be used when required.

17.4.2.2 Photovoltaic conversion

The direct conversion of sunlight into electricity is achieved by a process called the photovoltaic effect (photo = light, voltaic = electrical potential), discovered by a French physicist, Edmund Becquerel. The first solar cells were made of selenium in the 1880s with 1–2% efficiency (the percentage of available sunlight energy converted by the cell into electrical energy). By the mid-1950s Bell Telephone Labs had achieved 4% efficiency with silicon PV cells and strides due to growing energy demands, increasing environmental concerns and declining fossil fuel resources, which have resulted in falls in production costs to nearly 1/300 of what it was during the space programme of the mid-twentieth century (Alternative Energy Institute, 1999). The majority of power modules in use since 1955 are crystalline silicon or thin-film amorphous silicon. Satellites and other space applications predominantly use silicon or gallium arsenide. Other thin film materials include cadmium telluride (CdTe) and copper indium diselenide ($CuInSe_2$ or CIS).

When light shines on semiconducting materials, the photons (parcels of light energy) impart enough energy for some of the electrons to jump from a bound state to a free conducting state, leaving behind a hole which acts as a positive charge (see Figure 17.1). The hole moves by way of neighbouring electrons exchanging places with it. To make this useful, a cell is made from either two different semiconductors or the same but impregnated with different 'impurities' to create a junction (as close as possible to the surface where sunlight is absorbed) that separates positive and negative charges. This polarization of charges forms a voltage, and when connected to an external load it produces a direct current. The magnitude of this current is proportional to the intensity of the light. Cells wired together form a module, and modules wired together form a panel. A group of panels is called an array and several arrays form an array field.

The conversion efficiency of a solar cell is determined by losses arising from heat, reflection and non-absorption of photons by the semiconductor. Most solar cells actually work better in cold climates as thermally activated electron-hole recombination is reduced, and they also work equally well in cloudy conditions as in bright sunshine as both diffuse and direct radiation are absorbed. Thin films of poor quality silicon have been made into commercial low-cost multijunction solar cell modules with efficiencies reaching 20% (Anonymous, 1998a). A manufacturing firm in New Jersey produces a dual-junction solar PV cell for satellite applications with an efficiency rating of 25.3% (Canadian Association for Renewable Energies, 2000a). Spectrolab in California, which supplies PV cells for NASA, produces an average cell efficiency of 24.5%, and cells are on the production line with efficiencies of 27% (Canadian Association for Renewable Energies, 2000b).

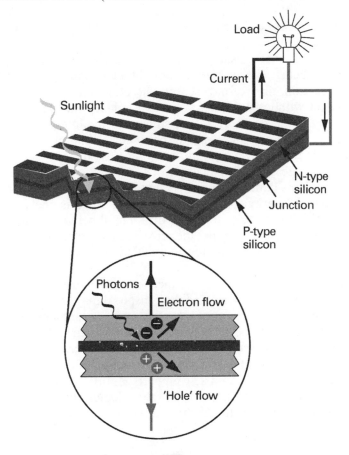

Figure 17.1 The photovoltaic effect in a solar cell
Source: *http://renewable.greenhouse.gov.au/technologies/*

A photovoltaic system has a number of advantages in that it has no moving parts and is non-polluting, quiet and safe, requires little maintenance and no supervision, and has a life of 20–30 years with low running costs. PV systems can be combined with other types of electric generators (for example, wind, hydro and diesel) to charge batteries and provide power on demand. Solar plants are modular and therefore easily scaled up or down and can be built up in much shorter times than a conventional power station. Solar arrays need not be intrusive or take up extra land space and can be incorporated into the external façade of buildings, e.g. solar cell roof tiles (Anonymous, 1998d) and a glass-louvre system that tracks the sun's movement to generate solar-electric power while reducing solar heat gain and glare in a building. A 9.5 kW system in Geneva generates up to 16 000 kWh of electricity per year (Anonymous, 1998b).

A total PV system includes a support structure, batteries (if remote), an inverter (which converts direct current power from the solar cells to alternating current) and wiring. An average home has more than enough roof area to produce enough solar electricity to supply all of its power needs. For a typical house of a daily average energy consumption of 10 kWh/day receiving an average 5 sun-hours per day, a 2 kW system would be needed to produce enough energy. This could be reduced with an energy-efficient home.

There are only two primary disadvantages to using solar power: the amount of available sunlight and cost. The amount of sunlight a location receives depends on its geographical location, time of day, season and cloud cover. Photovoltaic cells can work in daylight without sunshine, albeit at a reduced rate. Conservative calculations for the payback time of crystalline silicon PV modules varies from 2.5 to 5 years depending on how much sun a particular location receives. Future improvements in cell efficiency are projected to decrease the payback time to 1 to 2.5 years (Nijs et al., 1997). For other materials, such as amorphous silicon and polycrystalline thin films (CdTe, $CuInSe_2$), estimates are just more than a year or less for making up their energy cost (PV Power, 1999). Solar is unable to compete on a level playing field because of the price differential between solar and mains electricity. Environmentalists claim that the price of mains electricity is unrealistically low and does not take into account the hidden environmental costs (Planet Ark, 1999).

Environmental concerns of PV include their energy-intensive production, the use of scarce and/or toxic substances, and their disposal may lead to the release of hazardous substances. However, current recycling and environmental control technologies have been assessed to be sufficient to control wastes and hazardous emissions in production facilities, and progress in using thinner layers with better deposition processes or reuse would lead to further reduction of energy use and emissions and depletion of rare materials. Comparisons of life cycle assessments (LCA) of PV and non-PV power production technologies show that PV has significant environmental and health advantages over the latter (Nieuwlaar and Alsema, 1997).

17.4.2.3 Photoelectrochemical cells

By a process called photoelectrolysis, photons falling on an electrode in a photoelectric cell can produce a current that will split water into hydrogen and oxygen. Hydrogen can be transported more cheaply than electricity over large distances, stored indefinitely, and when it is burned as a fuel or used in a fuel cell to produce electricity, water is the only by-product (Hoagland, 1995).

A breakthrough in transparent cells of 10% efficiency has occurred in

Switzerland. A current is produced within a cell comprising an iodine-based elec-trolyte film trapped between two panes of glass. Both are coated with transparent tin oxide to form electrical contacts, and one has an additional transparent layer of the semiconductor titanium dioxide. Although it is of lower efficiency than normal solar panels, it is much cheaper and replacing normal windows with electrical windows results in a intrinsically low-cost source of electricity (Coghlan, 1995).

17.5 Other energy sources

17.5.1 Tidal power

Tidal energy is the conversion of gravitational energy and works on the same fun-damental principle as the water wheel except the difference in water elevation is caused by the difference between high and low tides. The technology involves building a dam, or barrage, across an estuary to block the incoming tide and/or the outgoing tide. Water pressure builds on one side of the dam due to a tidal change, and the water is channelled through a turbine to produce electricity.

Tidal energy is being harnessed in several countries around the world, particu-larly in areas of exceptionally high tides. A tidal range of at least seven metres is required for it to be economical. Tidal energy is constrained by the time lapse between changing tides, allowing it to generate a maximum of ten hours of elec-tricity per day. Although the operation and maintenance of a tidal power plant is low, the cost of the initial construction of the facility is very high. The alteration of the natural cycle of the tides may affect a shoreline as well as aquatic ecosys-tems. Pollution that enters a river upstream from the plant may be trapped in the basin, while the natural erosion and sedimentation pattern of the estuary may be altered.

An American developer has patented a scheme that would significantly reduce these limitations. For a 900 MW barrage that would normally block off a 954 km^2 basin of water, a new generator, built from two layers of concrete tanks in 30 metre deep water, would cover less than 1% of that area. The upper layer spans the height between the tides, and turbines are driven as water rushes in at high tide and out at low tide. The bottom tanks extend from the low water line to the seabed and at high tide water is admitted through a pump-turbine and pumped out at low tide, pro-ducing a net gain in energy due to the differences in weight of the water above the tanks (Hecht, 1995).

Despite tidal power's inability to replace conventional energy sources, it will not be dismissed in the near future. Britain, India and North Korea have planned to supplement their grid with this renewable energy source.

17.5.2 Geothermal energy

Several different types of geothermal energy exist. These include natural systems such as geysers and hot springs resulting from groundwater being heated by hot magma near the surface (1500 to 10 000 metres deep). Naturally occurring hot water and steam can be tapped by energy conversion technology to generate electricity or to produce hot water for direct use. Latent heat from dry steam, brine and hot dry rocks can be commercially used by drilling to between 3000 and 10 000 metres where the Earth's thermal gradient is sufficiently steep (Palz, 1978). In a hot dry rock geothermal system, water flowing down an injection well would seep through hot rocks that have been fractured and come up through a production well as superheated steam. Temperatures of 60°C are available in France from depths of 2000 metres and dry steam from deeper reservoirs in Italy and California are used for steam-powered plants. The availability of this resource is usually limited to geologically volatile areas, however a 1200 metre bore sunk into the Great Artesian Basin in Australia, extracts water at 98°C which is used in a power plant that produces 120 kW of electricity (Gilchrist, 1994). A typical geothermal site is illustrated in Figure 17.2.

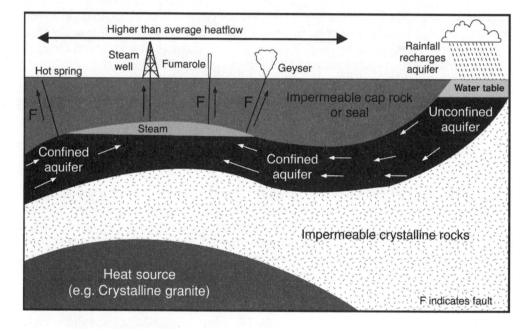

Figure 17.2 Simplified cross-section of the essential characteristics of a geothermal site
Source: http://renewable.greenhouse.gov.au/technologies/

Another form of geothermal energy involves using the heat stored in the Earth's surface. The ground tends to stay at a relatively constant temperature throughout the year, and this can be used to heat a building in winter and cool a building in summer. This form of energy can lessen the need for other power to maintain comfortable temperatures in buildings, but cannot be used to produce electricity.

17.5.3 Fuel cells

The fuel cell is an environmentally safe technology that produces electricity with no moving parts and very little pollution. It is an electrochemical device that consumes fuel, such as hydrogen, methanol or natural gas, to produce electricity at conversion efficiencies of up to 60% (or 80% if the waste heat is recovered). First invented in 1839, today hydrogen fuel cells provide much of the electric power for NASA's space shuttle, as well as for a wide variety of other uses from the smallest generator or engine to industrial and large commercial plants. Currently, fuel cells are being used in several automobile applications, and these non-polluting cars and buses are becoming commercially available in the US and Europe.

In a typical hydrogen fuel cell oxygen is taken up at the air electrode (positive cathode) and converted to negative ions that diffuse through a membrane and electrolyte to react with positive hydrogen ions at the anode (negative electrode) to produce water. Electric current is the result of electrons given up by the hydrogen flowing over to the air electrode via an external load. As an individual cell produces about one volt, any number of cells can be connected to form a fuel stack to produce a desired voltage and up to hundreds of megawatts.

A small amount of heat is released by the chemical reactions within the fuel cell, varying according to the size and design of the cell. Fuel cell design varies according to the power demands of a given system, and the operating temperatures that best suit that particular application. Higher operating temperatures allow the use of less pure hydrogen sources, as fuel cells are capable of chemically extracting hydrogen from a variety of fuels like methanol, and even fossil fuels like oil and natural gas.

Alkaline fuel cells used in NASA's space shuttle missions are expensive due to the use of platinum as a catalyst, but they can reach efficiencies of up to 70%. Phosphoric acid fuel cells (PAFC) have the widest commercial application. With operating temperatures around 200°C, the steam from the water produced within the cell can be collected and used (a process called cogeneration), raising the efficiency from 40% to 85%. PAFCs range in size from 5–200 kW and from 1–11 MW, and are mostly powered by natural gas. Proton exchange membrane (PEM) fuel cells (typical efficiencies of 30%) are the smallest and lightest and as they operate at only 95°C, they have faster start-up times, making them ideal for cars and trucks. Solid oxide fuel cells (SOFC) are suitable for large installations such as central electricity generation for utility companies and industrial production facilities. As these operate at temperatures of up to 980°C cogeneration is possible. Advantages of SOFC stacks are high fuel conversion efficiency of greater than 80% with heat recovery, low emissions, high power density (up to 1 MW/m^3), modular construction, fuel flexibility, very low noise and vibration levels and short installation lead times. Molten carbonate fuel cells are in competition with PAFCs for commercial applications like hotels and airport terminals. Using a molten mixture of carbonate salts to screen the ions within, and operating at around 650°C, this fuel cell design allows the use of less pure fuels, even coal- and oil-based hydrocarbon fuel sources.

Hydrogen fuel cell technology is limited by the high cost of hydrogen production (Canadian Association for Renewable Energies, 2000a). The most cost-efficient method currently employed is steam hydrocarbon reforming, where natural gas is chemically broken down by high temperature steam to release hydrogen. Electrolysis, where electricity is used to decompose water into hydrogen and oxygen, produces extremely pure hydrogen but is very energy intensive and expensive. Alternative sources of this electrical power include photoelectrochemical technology, facilitated either by parabolic solar collectors that concentrate the sun's energy, photovoltaic solar panel stacks, or hydroelectric power. Heat from collected solar energy can be used to form hydrogen directly from hydrogen bearing sources like water, natural gas, and organic biomass such as municipal and agricultural waste. Biomass sources can also be broken down biologically by various microbes to produce useable hydrogen. In photolysis, strains of blue-green algae are provided with certain temperature and light conditions so that the chlorophyll and enzymes within the algae can chemically split seawater into hydrogen and oxygen. Known as photobiological production, this is a very environmentally-friendly process, with no need for external energy, other than maintaining the conditions necessary to promote the process (Canadian Association for Renewable Energies, 2000a).

17.5.4 Nuclear fusion

Nuclear fusion is attractive as an energy source because of the virtually inexhaustible supply of fuel, minimal environmental impact, and its inherent safety. Opposite to nuclear fission, in which an atom of the heavy isotope Uranium-238 is split, fusion involves combining light atoms (isotopes of hydrogen – deuterium or tritium) to form the inert gas helium, while releasing an enormous amount of energy.

The fundamental problem with nuclear fusion is that the fuel must be raised to over one hundred million degrees to form a fully ionized state called a 'plasma'. At these temperatures the plasma can only be contained by strong magnetic fields, typically 100 000 times the Earth's magnetic field, or by inertial confinement which requires powerful lasers or high energy particle beams to compress the fusion fuel. Another fundamental problem with hot fusion is the lack of a net energy gain due to the demanding requirements involved with burning plasma. If the process could be brought about at room temperatures, the complexity that now prevents the generation of power based on nuclear fusion would disappear. Hence the excitement associated with the announcement of a 'cold fusion breakthrough' at the University of Utah, which unfortunately has not been replicated.

17.6 Conclusion

A fundamental process on which human existence depends is the transformation of energy into another form. Generally speaking conversion efficiencies are quite poor, however those of renewable energy sources are often far superior to that of fossil fuel generating systems. The maximum efficiency attained by a coal-fired power station is approximately 36–38% (although cogeneration can increase this to 60–80%), whereas hydro is greater than 90%, wind is approximately 40%, fuel cells are 60–80% and biomass is about 80%. The comparison is more dramatic when one considers, apart from the limited fuel supply and associated pollution of the generated energy, that fossil fuel originally came from a photosynthetic process which is far less efficient than a solar cell and used only 1% of the available sunlight reaching the Earth's surface.

The developing world, which accounts for a large portion of the projected increases in electricity demand over the next 20 years, is considered to be one of the biggest arenas for renewable energy. Providing renewable energy technologies to developing countries is the best way to deal with local pollution problems, minimizing the emission of greenhouse gases and combating global warming. Wind power, solar power, biogas and mini-hydro are already proving successful in various regions. In industrialized countries, the desire for clean and sustainable energy sources is causing the renewable energy market to expand rapidly. Renewable energy already accounts for 6% of total energy in Europe and this is targeted to double by 2010 (Canadian Association for Renewable Energies, 2000c). In the US, national surveys show that well over half of consumers are willing to pay more for green power.

Successful implementation requires impetus at the government and utilities level entailing programmes that include promotion and subsidies, tariff settings, regulations, rebate schemes and premium payback rates into the grid. Community education is also necessary to dispel widespread notions that renewable energy systems are unreliable and inefficient and to ensure that renewable energy can cost less in certain circumstances than conventional power.

References and bibliography

Alternative Energy Institute website (1999). <http://www.altenergy.org/>

Anonymous (1998a). 'Solar Two World Records'. *Australian Energy News*, Issue 7, March, pp. 6–8.

Anonymous (1998b). 'Shadovoltaics'. *Australian Energy News*, Issue 10, December. <http://www.isr.gov.au/aen/aen10/10shado.html>

Anonymous (1998c). 'Surge of Interest in Solar Thermal Power'. *Australian Energy News*, Issue 8, June.

Anonymous (1998d). 'Solar Tiles'. *Australian Energy News*, Issue 8, June. <http://www.isr.gov.au/aen/aen8/8tiles.html>

Australian Renewable Energy (1999). 'Wind Energy Technologies'. <http://renewable.greenhouse.gov.au/technologies/wind/wind.html>

Bickers, C. (1999). 'The Next Wave'. *Far Eastern Economic Review*, vol. 162, no. 2, p. 74. <http://db.webhk.com/9901_14/p71tech.html>

CADDET (1997). *CADDET Renewable Energy Newsletter*, Issue no. 2, June, pp. 18–19.

Canadian Association for Renewable Energies (1999). *Trends in Renewable Energies*, 8 February. <http://www.renewables.ca>

Canadian Association for Renewable Energies (2000a). *Trends in Renewable Energies*, Issue no. 116, 31 January – 4 February. <http://www.renewables.ca>

Canadian Association for Renewable Energies (2000b). *Trends in Renewable Energies*, Issue no. 126, 10–14 April. <http://www.renewables.ca>

Canadian Association for Renewable Energies (2000c). *Trends in Renewable Energies*, Issue no. 125, 3–7 April. <http://www.renewables.ca>

Coghlan, A. (1993). 'Algal Power Gives a Clean Burn'. *New Scientist*, vol. 137, no. 1856, p. 18.

Coghlan, A. (1995). 'Sunnier Outlook for Powerful Windows'. *New Scientist*, vol. 145, no. 1962, p. 22.

Denniss, T. (2000). 'Wave Energy System using Parabolic Focussing Walls'. *CADDET Reporter*, no. 29, April, p. 1.

Dovers, S. (1994). *Sustainable Energy Systems:* Pathways for Australian Energy Reform. Cambridge University Press.

Elliott, D. (1997). *Energy, Society and Environment: Technology for a Sustainable Future*. Routledge.

Endo, S. (1995). 'The Mighty Whale that Rules the Waves'. *New Scientist*, vol. 148, no. 2005, p. 42.

Etherington, D. (1995). 'How many Coconuts to the Kilometre?'. *Energy Focus*, vol. 45, March, pp. 2–4.

Freere, P. (1995). 'Energy from the Wind'. *Solar Progress*, vol. 16, no. 4, pp. 7–9.

Geary, J. (1995). 'Plastic Cables Pull Power from the Sea'. *New Scientist*, vol. 145, no. 1960, p. 19.

Gilchrist, G. (1994). *The Big Switch: Clean Energy for the Twenty-First Century*. Allen and Unwin.

Gilchrist, G. (1995). 'Our Solar Battery: A Powerful First'. *Sydney Morning Herald*, 18 November, pp. 1–2.

Hamakawa, Y. (1987). 'Photovoltaic Power'. *Scientific American*, vol. 256, no. 4, pp. 77–82.

Hecht, J. (1995). 'High Time for Compact Tidal Power?'. *New Scientist*, vol. 145, no. 1964, p. 20.

Herzog, P. (1997). *Energy-Efficient Operation of Commercial Buildings: Redefining the Energy Manager's Job*. McGraw-Hill.

Hoagland, W. (1995). 'Solar Energy'. *Scientific American*, vol. 273, no. 3, pp. 136–39.

Hodgson, P.E. (1997). *Energy and Environment*. Bowerdean Publishing.

Integrity Research Institute (2000). Press Release: Alternative Energy Institute, 15 May. <http://www.altenergy.org/News/news.html>

Kaneff, S. (1994). 'Solar Generator 3: ANU's 400 sq.m Paraboloidal Concentrator'. *Solar Progress*, vol. 15, no. 3, pp. 4–5.

Luzzi, A. (1996). 'Don't Forget Solar Chemistry'. *Solar Progress*, vol. 17, no. 1, p. 5.

McIver, A.D., Freere, P. and Holmes, D.G. (1995). 'Grid Connection of a Variable Speed Wind Turbine Using a Matrix Converter'. *Solar Progress*, vol. 15, no. 4, p. 30.

Moss, K.J. (1997). *Energy Management and Operating Costs in Buildings*. E. & F.N. Spon.

Nieuwlaar, E. and Alsema, E. (1997). 'Environmental Aspects of PV Power Systems'. <wire1.ises.org/wire/Publications/>

Nijs, J., Mertens, R., Van Overstraeten, R., Szlufcik, J., Hukin, D. and Frisson, L. (1997). 'Energy Payback Time of Crystalline Silicon Solar Modules'. In *Advances in Solar Energy* (K. Boer, ed.), ASES, Boulder, CO, USA, vol. 11, pp. 291–328.

Page, D.I. and Legerton, M. (1996). 'Wind Energy Implementation During 1996'. *CADDET Renewable Energy*. <www.caddet-re.org/html/397art6.htm>

Palz, W. (1978). *Solar Electricity: An Economic Approach to Solar Energy*. Butterworths.

Planet Ark website (1999). <http://www.planet.ark.com.au/>

PV Power (1999). 'PV FAQ'. <http://www.pvpower.com/pvfaq.html>

Rasmussen, A. (1995). 'A Brief History of Solar Energy Technology'. *Solar Progress*, vol. 15, no. 4, pp. 22–23.

Ristinen, R.A. and Kraushaar, J.J. (1999). *Energy and the Environment*. John Wiley & Sons.

Singh, D. (1995). 'The Tyson Turbine: Another Remote Area Power Supply'. *Solar Progress*, vol. 16, no. 3, pp. 10–12.

Thomas, R. (1999). *Environmental Design: An Introduction for Architects and Engineers* (2nd edition). E. & F.N. Spon.

Tuluca, A. (1997). *Energy Efficient Design and Construction for Commercial Buildings*. McGraw-Hill.

Venendaal, R. and van Haren, P. (2000). 'Alternative Fuels in Electric Power Generating Plants'. *CADDET Energy Efficiency Newsletter*, no. 1, March, p. 24.

Weinberg, C.J. and Williams, R.H. (1990). 'Energy from the Sun'. *Scientific American*, vol. 263, no. 3, pp. 99–106.

Wind Energy Training Centre (2000). 'Power from the Waves'. March. <http://www.iesd.dmu.ac.uk/~slb/wcwave.html>

Yan, J., Alvfors, P., Eidensten, L. and Svedberg, G. (1997). 'A Future for Biomass'. *Mechanical Engineering Magazine Online*. <http://www.memagazine.org/backissue/october97/features/biomass/html>

18

Energy regulation and policy

18.1 Introduction

The world's transport systems run largely on oil, as does much of mechanized agriculture. Similarly oil production is essential to the continuation of other economic activities and the general support of daily life. According to Fleay (1995), oil production in the major producing regions of the world has reached its peak and is now beginning to decline. The giant oil reserve, the Arabian Gulf region that contains two-thirds of world oil reserves, is expected to peak by the year 2005 (formerly predicted to peak by the year 2020). Overall, oil production in the world is likely to peak some ten years earlier than previously expected.

Following the first oil crisis in the 1970s, energy policy internationally was focused on world oil reserves and reliable supply at reasonable prices. Much research was undertaken and government policy was mainly concerned with conserving non-renewable energy resources and their replacement with renewable alternatives. As time passed, the dark age of the oil crisis faded and government attention was diverted to other issues. Energy policies, therefore, were given a lower priority on the national and international agenda. In the 1990s, with the discovery of the Antarctic ozone hole and the potential threat to global climate change, energy policy has crept back onto the political agenda and is now receiving attention worldwide.

Energy, which is closely linked with environmental degradation and resource usage, has become the main focus of discussion in seeking new solutions. Strong evidence of the warming of the Earth's surface as a result of an unprecedented increase in the atmospheric concentration of greenhouse gases has required international co-operation to reduce harmful emissions. The main objective is to raise international awareness towards solving environmental problems now rather than leave them for future generations to overcome.

While international agreements and targets can be influential, regulation and policy implementation is within the domain of individual countries. Common strategies include incentive schemes, voluntary standards, mandatory standards and energy labelling. The processes of construction and operation of buildings play an important role in the management of energy use, since buildings are major contributors to energy consumption and hence drive energy production and greenhouse gas emission.

Sustainability is a global concern. It is the prime target of all governments to take up the environmental challenge without adversely affecting economic growth. Energy services are fundamental to living standards and to the conduct of economic activities. Many countries, while confronted with environmental problems that are closely linked with energy consumption, have established long-term policies to tackle both the environmental and energy issues. The aims of energy policy are to secure continuous energy supply, pursue development of the energy exports sector and achieve energy efficiency in all sectors of society. If energy supplies are inadequate or delivered inefficiently, this will represent a serious constraint on economic growth and impact on future living standards.

18.2 Incentive schemes

In countries without effective energy-efficiency programmes for their buildings, current building energy use trends are (or should be) cause for concern. The building sector consumes roughly one-third of the final energy used in most countries, and it absorbs an even more significant share of electricity. In developed countries, buildings account for 57% of electricity use, 31% in the residential sector and 26% in the commercial sector. In less-developed countries, buildings account for 38% of electricity consumption, and electricity use in this sector grew more than 11% per year over the 1980s (Janda and Busch, 1994). Electricity use in commercial buildings is driving peak demand in the US, Japan and in some of the wealthier less-developed countries. As people raise their standards of living and services, electricity use in buildings is expected to continue to increase, especially in the commercial sector.

People living in industrialized countries are a relatively small percentage of the world's population, but they consume a disproportionate share of the total energy produced. For example, the US has only 5% of the world's population but uses approximately 25% of the total energy consumed on the planet (Botkin and Keller, 1995). There is a direct relationship between a country's standard of living (as measured by gross national product) and energy consumption per capita. This relationship needs to reflect more efficient use of energy resources rather than a continuous increase in total energy consumed.

Incentive schemes have been introduced worldwide to encourage solutions to be found to reduce the demand for energy. In a sense this is related to the predominant use of non-renewable energy and the knowledge that conservation of limited resource stocks must occur. Incentive schemes aim to reward consumers, either financially or through market advantage, if they are able to demonstrate success in reducing their demand for energy.

Incentive schemes function best in a free market economy. They usually involve some form of government intervention, but can become self-sustaining and market-driven. Public awareness and promotion are key attributes of a successful incentive scheme. Some examples include rebates on insulation or solar hot water, recycling, subsidized technologies, promotion of sustainable (green) energy production, and showcasing exemplary behaviour in magazines and websites.

However, successful incentive schemes require some form of benefit to be received. Where there is little incentive to change current behaviour, performance standards need to be introduced to regulate activity. The construction industry is one example where incentives are useful but often insufficient to make a large difference to total energy consumption patterns.

18.3 Voluntary standards

National and international initiatives suggest that the question about energy standards for the future is not whether governments will regulate building design to limit energy consumption, but in what ways and by how much. There is considerable evidence worldwide of energy standards that have been implemented to control the built environment, and more are being added each year (Janda and Busch, 1994).

To broadly characterize the status of energy standards for buildings, Janda and Busch (1994) combined previously published information with results of a worldwide survey across 57 countries. Thirteen of the countries for which information was gathered had no energy standards for any building sector; four countries had standards only for the residential sector; nine countries developed standards exclusively for non-residential buildings; and 31 countries had standards for both. At the national level, 21 countries had mandatory energy standards for at least one building sector, and 11 had voluntary standards. Six more countries have proposed but not yet adopted energy standards; half of these are non-residential standards and the rest are for both residential and commercial or all buildings.

Voluntary standards are common worldwide and represent an attempt by government agencies, private associations and professional advisers to agree on good practice. Voluntary standards work on the principle of identifying benchmarks which designers should aim for in the development of new buildings or major refurbishment. They can also apply to existing buildings and the management of operational performance.

Voluntary standards, however, are generally regarded as weak in bringing about widespread change. While they represent non-government policy initiatives, by their nature they are implemented by agreement and are non-enforceable. Voluntary standards often form an interim control mechanism until more stringent mandatory standards are shown to be necessary, perhaps due to the inability of the marketplace to correct and regulate itself.

Even a well-designed mandatory standard will not save energy if it is not followed. Implementing an energy standard involves a network of social systems and human interactions that stretches from the bureaucrats assigned to administer the standard to the tradespeople who install the chosen solutions. Any person or group involved in a building's development, design, and construction process can affect its final energy use, so there are an almost unlimited number of opportunities for the building to comply with or deviate from the standard's recommendations. The power of the implementing agency, the level of training provided, and the effectiveness of compliance mechanisms are all important indicators of the extent to which the standard is likely to be followed (Janda and Busch, 1994).

18.4 Mandatory standards

Where market failure exists and affects the introduction of technologies and processes that improve energy performance, government intervention is necessitated. Such action could set performance requirements for new buildings, refurbished buildings and existing buildings offered for sale. This process must take account of both embodied energy and operational energy attributes (yet not be limited to energy alone) and establish clear boundaries which developers, designers and owners must work within. Such an approach will change consumer behaviour, bring about progressive improvement in the quality of a nation's building stock, and encourage innovation.

Although the environmental performance of commercial buildings incorporates many issues, energy is an obvious one. In most countries the vast majority of energy used in commercial buildings is electricity. The low price of electrical energy is a major reason why building owners and tenants are not overly concerned about further capital spending to reduce energy consumption. Studies by the European Union found that the external costs associated with coal-fired electricity were in the vicinity of US$0.20 per kWh (cited in Diesendorf, 1996) and were not built into the consumer price. If pricing policy were set to reflect actual social/environmental costs, in many countries this would lead to at least a threefold increase in the price (not cost) of electricity. However, even a doubling of price would be politically untenable, and all-pervasive throughout society.

Financial incentives operating in a competitive market without government intervention are obviously favoured. However, issues of environmental protection and pursuit of sustainable development ideals are important, and may be in conflict with market-driven investment goals. As environmental problems become more pronounced there must be an increase in planning controls that avert 'bad' behaviour.

The car industry is an interesting analogy. While fuel consumption is an issue with many, people still purchase fuel-inefficient vehicles and frequently choose low purchase options in lieu of cars with long-term reliability or durability benefits. Much the same behaviour is observed with commercial buildings. However, if the price of petrol was to rise significantly (due to crude oil price rises by producers), then we expect a reduction in total fuel consumed. Recent evidence has shown we would be disappointed. In fact, it would probably take a significant rise in petrol prices (or fuel scarcity) to engender any real behavioural change. The consumer fall-out would also be significant, and it would be a brave politician that would signal a doubling of petrol prices, as a means of environmental conservation, with a view to reducing the amount of cars on the road.

But government intervention in the performance standards of new cars, particularly if this did not translate into significant increases in capital cost, would result in more efficient cars using less fuel per kilometre. This would be likely to encourage greater use of these efficient cars, so that the total fuel consumed could end up much as before. More cars or more journeys multiplied by less fuel may not result in an overall social benefit; in fact it may cause more congestion on roads, parking shortages, more pollution and accidents.

It is also interesting to consider the issue of innovation. The car analogy can be further extended. Back in the 1950s and 1960s safety was considered an emerging social issue. Many governments intervened by requiring that all cars be fitted with front and rear seat belts. In effect, they set minimum standards. In the years that followed, however, we saw the introduction of more safety features as a result of market expectations and competition. Innovations like hazard lights, roll bars, side impact resistance bars, ABS and air bags are but a few examples. Perhaps the increased awareness of safety, through government intervention and through road safety awareness programmes, educated the marketplace to want and indeed expect high levels of safety in their cars. The majority of new cars sold to consumers are now of very high standard, well above what might be considered as a minimum performance requirement.

Could the same thing happen to buildings? The introduction of energy-saving technologies and processes would result in lower energy costs, which might mean that building owners or tenants might translate that saving back into higher usage. Contemporary literature on facility management, however, indicates that businesses are looking for ways to become more competitive, and this usually means strategic reorganization of work practices, improvement of worker productivity and the control of costs. In this economic environment it may be equally likely that savings in energy and pursuit of low maintenance alternatives will be targeted so that profit levels can be maximized. The concept of buildings being lit or air conditioned all night simply because energy was cheap is hard to imagine. Yet there is no reason why innovation should be inhibited, as the marketplace will demand new solutions if awareness of the issues and advantages are promoted properly.

There are a large number of examples worldwide where governments (federal, state and local) have introduced schemes to encourage operational-saving behaviour in the built environment. What is clear is that the wider community is the main beneficiary of investment in energy-saving measures. The aim is to reduce total energy demand, which includes energy embodied in the manufacture of materials, and make buildings more sustainable through lower use of resources and the associated impact on the environment. Issues of enhanced greenhouse effect, climate change and exponential use of non-renewal resources require that the government act to protect society from itself.

It is also clear that the construction industry is the largest consumer of resources, and hence has the most potential for savings. Issues such as refurbishment, adaptive reuse, recycling, natural lighting, natural ventilation and proper design take on greater importance. Designers need to be well educated about ways to improve value in buildings, particularly at early stages in the procurement process (Best and de Valence, 1999). Short-sightedness and ignorance of available solutions is considered the main impediment.

Designers need to meet higher targets of performance, and therefore find building solutions which deliver social benefit. Why will developers and short-term investors co-operate? Government intervention is necessary to correct the inherent market failure caused by investor-centred behaviour. This can be achieved in part by requiring that new buildings meet minimum energy performance standards (i.e. exhibit total energy consumption within legislated maximum limits). In time this can be extended to refurbished buildings and existing buildings offered for sale.

Such a strategy requires research into what level to set the benchmarks, and it is important that benchmarks are not pitched so low that they have little effect. Energy labelling systems that can communicate the performance of buildings to the public, including prospective tenants, will eventually become a key factor in determination of property values and rent levels as consumers become more aware of environmental performance and associated financial benefits.

18.5 Energy labelling

Evidence from around the world suggests clearly that in order to improve industry performance it is necessary to set a framework within which the marketplace can be allowed to find its own solutions based on simple competitive advantage. However, this framework requires government intervention in the early stages to steer the various stakeholders in desired directions.

One of the major obstacles to implementation of assessment techniques and advanced construction solutions is making the motive for such action relevant to the speculative investor or developer. While it is possible to set up a framework within the public sector through policy change, the success of any action lies in its take-up rate by the private sector. It is generally accepted that an incentive approach is preferred to a legislative one, but to be effective it must relate to all sectors of the industry and at all levels of investment.

The following recommendations are proposed as one method to achieve this framework within the short term. While they focus on operational energy performance, in time the inclusion of embodied energy will ensure that issues of maintenance, replacement and recycling are part of the decision process to arrive at efficient design solutions. However, the strength of the approach lies in the ability of the marketplace to view the recommendations as a means of improving financial return in property transactions through a changed consumer focus and a better understanding of energy performance and resource efficiency.

It is therefore recommended to:

1. *Establish a five-star energy rating system for all building projects.* Improvements in operational energy performance are a key factor in the reduction of greenhouse gas emissions. The star rating system has the advantage of being easily understood by the community, and in essence is little different to that used commonly on white-good products such as refrigerators and dishwashers. The rating system is best structured on a five-point scale (five stars equal excellent energy efficiency) in increments of half stars. While the calculations behind the end result may be complex, the simple presentation will enable tenants and purchasers of buildings to correctly interpret the impact of operating performance and, therefore, incorporate such information into their financial decisions.

2. *Ensure all building approvals are conditional upon achievement of at least three-star operational energy performance.* If improvements in building performance are to be realized, minimum standards of compliance are necessary. It is suggested that a three-star rating may be a suitable benchmark initially. This can be adjusted upwards in the future as governments wish to exercise more stringent control over development. The benchmark is a performance standard, not a design solution, and therefore it will encourage innovation in the industry and enable delivery of necessary performance at minimum total (life) cost.

3. *Amend building codes to include achievement of a three-star energy rating as an essential minimum performance standard.* To be effective, this standard should apply to all classes of buildings (both new work and renovation). Building codes are the best instrument to do this, and local authorities will therefore have the responsibility to ensure compliance. Planning approval must be conditional on the minimum energy rating being achieved. The changes necessary to building codes are probably quite minimal.

4. *Develop tools to assist professional advisers to make and interrogate star rating calculations.* A common methodology needs to be developed to measure star ratings. Software tools can be used by professional advisers, contractors, individuals and local authorities to assess designs and interrogate the results so that the best balance of performance and cost can be realized. These tools need to be made available in the marketplace at reasonable cost. The methodology for calculation must be agreed so that results are repeatable. In time, automatic calculation linked to three-dimensional CAD models will provide productivity improvements in practice.

5. *Provide research funding to assist development of benchmarks for assessing energy performance and star ratings in a range of environmental contexts.* Different climatic zones will give rise to different construction solutions, but the star rating system must be able to apply to all conditions in an equitable fashion. Further research is necessary to identify suitable benchmarks for various classes of buildings and to link them to the five-star scale. Factors that will impact on performance include orientation, aspect, glazing, thermal resistance, material selection, insulation, HVAC decisions and the like. The calculations are complex and need to be suitably modelled by software if the approach is to work in practice.

6. *Launch an effective promotion and education campaign to encourage pursuit of and consumer demand for energy-efficient solutions.* It is critical that an appropriate campaign of public information be launched to introduce the new requirements and to instil in consumers the importance of each star rating increment in terms of annual operating cost. Continuing professional education for client advisers is also necessary. In the longer term it is the marketplace itself that must drive the search for more efficient design solutions, and for this to occur consumers must be properly informed.

7. *Encourage existing building owners to use star ratings when advertising for tenants or purchasers.* Coupled with promotion and education campaigns, it is important to work with property agents to ensure that the energy performance rating is made clear to potential consumers. In much the same way as

food needs to be clearly labelled, so too buildings must have compulsory energy labelling. In the short term this will apply to new or renovated buildings, but over time this will naturally spread to existing buildings. Market forces may encourage building owners to prematurely undertake energy classification to improve their ability to get the best market price.

8. *In the longer term, include embodied energy considerations in the calculation of star ratings*. Operational energy deals only with annual energy demands. The original construction process (including the whole chain of mining, manufacturing and delivering installed materials) also consumes energy that becomes embodied in the finished components and systems. In the same way, maintenance and replacement activities consume energy. Therefore a building that has low operational energy may achieve this at the expense of high embodied energy, and overall little may be gained if operational energy is considered in isolation. Total energy (embodied plus operating) is therefore a better basis for calculation of star ratings, but this level of sophistication will require greater research and development before it can be used in a practical sense.

In addition, for public projects, the following recommendations should be implemented as a means of raising the profile of whole-of-life assessment and to encourage better decision-making processes during the design stage. It is expected that the systems that professional advisers will develop and use for public projects will also be used for private sector work. It is therefore further recommended to:

9. *Introduce mandatory life-cost assessment into the design process of every major new public sector project*. The public sector has a role to play in encouraging the construction industry to adopt better practices. The investigation of operational performance through life-cost studies for all new public sector projects should be mandatory. This process will enable building performance to be further improved and encourage the development of new design solutions that maximize social benefit over a long time horizon. Governments are not only major clients for the industry but also long-term occupiers of space, and therefore market intervention of this type can make a significant difference.

10. *Merge capital and operating budgets for all government agencies*. The principle of life-cost assessment requires the ability to balance initial and recurrent expenditure to achieve the best result. In cases where capital budgets are isolated from maintenance budgets, the whole concept of life-cost assessment is pointless. Private sector clients have more flexibility to adjust to this practice than government agencies, and therefore it is a priority that policy be altered to enable holistic construction budgets to be introduced.

The previous recommendations, if implemented, will lead to a significant improvement in the performance of the construction industry both in terms of the development of more efficient energy systems and the better utilization of national resources.

18.6 Conclusion

Energy regulation and policy is an important part of sustainable development and one that must surely require further consideration in the years ahead. The continuing demand on non-renewable energy resources, the overuse of renewable resources, and worsening of pollution, ozone depletion and global warming necessitate governments worldwide to control development and ensure that it is more sustainable. The success or otherwise of this endeavour will significantly dictate the living standards of future generations.

So what is the solution? Maybe there is no simple answer, but rather a complex array of initiatives and controls that will collectively create awareness of sustainable development and assist in educating the marketplace to expect higher performance. A strategy that may be worth considering is the mandatory preparation of a 'sustainability plan' as part of the building application process. This instrument could address key environmental performance indicators, such as operating energy efficiency, as a prerequisite to planning consent. The BREEAM assessment system, developed in the UK by the Building Research Establishment but now adapted to Hong Kong and Canada, may also form a suitable methodology for application to commercial buildings in other countries (<www.bre.co.uk/bre/breeam2.html>).

All signs point to the conclusion that energy standards, particularly for non-residential buildings, will play an increasingly significant role in the future of national and possibly international energy-efficiency policies. While energy standards for buildings have been developed in at least one-third of the world's countries, the other two-thirds have few ways of learning about the existence of information on this topic, and all countries currently face barriers to accessing information (Janda and Busch, 1994).

Discounting with rates well above the weighted cost of capital have encouraged decision-makers to value present actions much more highly than those that occur in the future. In the late 1980s and throughout the 1990s this has changed due to a greater recognition that high discount rates are in direct opposition with sustainable development goals. Cost-benefit analysis and life-costing are effective tools that can assist in the economic justification of decisions over a number of years. In combination with low discount rates (such as 2% to 8%), decision-makers are realizing that proper decisions can be made only if all costs are considered, not just those that involve initial construction.

References and bibliography

Best, R. (1996a). 'Controlling Energy Demand in Buildings: Submission to the Task Force, White Paper on Sustainable National Energy Policy'. Unpublished paper submitted to Department of Primary Industries and Energy, University of Technology, Sydney.

Best, R. (1996b). 'The Building Energy Code of Australia: An Opportunity Missed'. In proceedings of CIB Commission Meetings & Presentation, RMIT, Melbourne, February.

Best, R. and de Valence, G. (1999). *Building in Value: Pre-design Issues*. Arnold Publishers.

Botkin, D. and Keller, E. (1995). *Environmental Science: Earth as a Living Planet*. John Wiley & Sons.

Bureau of Industry Economics (1994). *Energy Labelling and Standards*. AGPS, Canberra.

Department of Primary Industries and Energy (1995). *National Sustainable Energy Policy: A Discussion Paper*. AGPS, Canberra.

Diesendorf, M. (1996). 'How can a "Competitive" Market for Electricity be made Compatible with the Reduction of Greenhouse Gas Emissions?'. *Ecological Economics*, no. 7, pp. 33–48.

Editor (1994a). 'Energy Performance Standards for Industrial and Commercial Equipment: The North American Experience'. *Australian Energy Management News*, Issue no. 23, pp. 22, 27.

Editor (1994b). 'Fuel Consumption Labelling of New Passenger Vehicles'. *Australian Energy Management News*, Issue no. 25, pp. 26–27.

Elliott, D. (1997). *Energy, Society and Environment: Technology for a Sustainable Future*. Routledge.

Energy Research and Development Corporation (1994). *Research and Development Plan 1994/95 to 1998/99*. ERDC.

Fleay, B.J. (1995). *The Decline of the Age of Oil*. Pluto Press.

Herzog, P. (1997). *Energy-Efficient Operation of Commercial Buildings: Redefining the Energy Manager's Job*. McGraw-Hill.

Hodgson, P.E. (1997). *Energy and Environment*. Bowerdean Publishing.

Janda, K.B. and Busch, J.F. (1994). 'Worldwide Status of Energy Standards for Buildings'. *Energy*, vol. 19, no. 1, pp. 27–44.

Krause, F., Bach, W. and Kooney, J. (1990). *Energy Policy in the Greenhouse*. Earthscan Publications.

Moss, K.J. (1997). *Energy Management and Operating Costs in Buildings*. E. & F.N. Spon.

Productivity Commission (1999). *The Environmental Performance of Commercial Buildings: Research Report*. AusInfo, Canberra.

Ristinen, R.A. and Kraushaar, J.J. (1999). *Energy and the Environment*. John Wiley & Sons.

Thomas, R. (1999). *Environmental Design: An Introduction for Architects and Engineers* (2nd edition). E. & F.N. Spon.

Tuluca, A. (1997). *Energy Efficient Design and Construction for Commercial Buildings*. McGraw-Hill.

PART 7

Life-cost studies

Once development feasibility is determined, a process of design and optimization follow to translate expectations into reality. The analysis of these more detailed decisions is usually made in economic terms against a fixed performance benchmark. It is essentially a 'value for money' decision process and once again involves striking an effective balance. The performance benchmark, at this stage enshrined in the client brief, comprises functionality, quality, amenity, environmental impact and other key criteria. Cost, which can be objectively measured to reflect the choices being made, is deliberately excluded as a performance criterion.

Design solutions that exceed set performance benchmarks usually come at a higher cost. Optimization of value for money normally commences by identifying the required performance benchmarks and looking for ways to deliver them at lower total cost. Increases above set benchmarks will normally be accepted only if the extra cost represents good value.

Cost is not limited to capital cost items incurred in the realization of the development, but must also take into account operating expenditure and financial liabilities. The combination of capital, operating and financial expenses is called life-cost. For projects that span more than several years, life-costs should be presented as a cash flow and discounted so that cash flows further away from the present are given less weight.

Value for money for any particular component of a project, or for the project as a whole, is identified as total performance divided by discounted life-cost. Performance is presented as a value score derived by multiplication of individual criteria weights with assessed achievement rankings. The higher the value score the higher is the performance level, while lower discounted life-costs are preferred. A ratio is produced and used to rank options for final decision. The higher the ratio, the better value for money the option represents, given that performance does not fall below prescribed benchmarks. Using this approach, improvements to facilities either during their design or after occupation can be objectively evaluated. It is incorrect to assess design based solely on minimum cost, just as it is inappropriate to treat project sustainability as being the same thing as project feasibility.

Discounting is nevertheless a fundamental component of design optimization and it is important that the technique is properly applied to avoid decisions that are biased towards present (or future) generations. Discounting is relevant to any financial comparison that embodies a time dimension. However, discounting should be applied only to items that are tangible, or in other words, cash flows that represent real monetary transfers. All other considerations, including those of an environmental nature, are judged as part of the value score.

The concept of design optimization can be extended for income-producing buildings to embrace potential productivity gains from occupant activities within the completed facility. Life-costs are held to include occupancy costs, also known as functional-use costs, such as staffing and other business-related expenditure. This creates a further level of complexity, particularly as occupancy costs can make conventional building costs appear insignificant over long time horizons. However, it does raise an important issue. If cost is to form part of the decision process, then it must include all cost items, not just those that are the easiest to identify.

Although the concept of considering all the costs of a project may seem common sense, in current practice decisions are made largely, and unfortunately, on the basis of capital costs. This results in value for money being viewed in a narrow context and limits the possibility of achieving efficiencies in the deployment of resources within society. While value assessment must integrate all relevant factors, including functionality, aesthetics, environmental impact and financial return, operating costs remain a significant part of any asset decision. The design professions must be conscious of the importance of operating costs and seek out solutions in their practice that will not ultimately become financial burdens for their clients or others in the community. The ability to deliver value for money on projects whilst concentrating solely on capital cost issues remains a contemporary fallacy.

The assessment of detailed design decisions is therefore a blend of traditional discounted cash flow analysis and value management techniques capable of making the trade-off between performance (maximum utility) and cost (minimum resource input) transparent. Design solutions given the most favour are those that can deliver performance at the lowest unit cost. This is a useful approach for making informed decisions about complex problems and a vital step in the pursuit of sustainable development.

The chapters in this part focus on the identification of value for money in the context of design development. Chapter 19 examines the techniques of life-cost planning and life-cost analysis and the key issues that are relevant when evaluating decisions. Chapter 20 investigates the discounting philosophy and how to correctly apply it. Chapter 21 discusses occupancy costs and the significance of productivity improvements that can be introduced by some thoughtful environmental design.

It is important that financial evaluation is not allowed to dominate decisions. It is equally important that value for money is not ignored. Projects that exhibit sustainable characteristics are those that deliver required social infrastructure through the efficient use of resources (i.e. capital wealth and environmental wealth).

19

Life-cost planning and analysis

19.1 Introduction

Commonly the measurement of costs is undertaken on a capital cost basis. Projects are analysed in respect of their likely construction costs, to which might be added land purchase, professional fees, furnishings and cost escalation to the end of the construction period. Budgets similarly address initial costs, and the planning and control processes they foreshadow normally do not extend beyond hand-over.

The need to look further into a project's life than merely its design or construction is a trivially obvious idea, in that all costs from an investment decision are relevant to that decision (Ashworth, 1999). The total cost approach takes into account both capital and operating costs so that more effective decisions can be made.

Terminology in common service to describe the total cost approach includes life cycle cost, life-cost, recurrent cost, cost-in-use, operational cost, running cost, ultimate cost and terotechnology. The proliferation of terms has resulted from the progressive debate and redefinition of the total cost approach in a number of countries, principally the UK and the US.

Life cycle cost is now the most popular term. The technique (life cycle costing) is generally defined as the economic assessment of competing design alternatives, considering all significant costs of ownership over the economic life of each alternative, expressed in equivalent dollars (Dell'Isola and Kirk, 1995a). The technique takes account of both initial design and acquisition costs and subsequent running costs and is measured over the asset's 'effective' life, generally considered as economic or operating life. Although many definitions exist, the majority excludes matters of revenue generated by the asset, other than taxation concessions and salvage value.

Life-cost is an alternative descriptor that appears to have independent origins in Australia, New Zealand and Canada. Life-cost studies can be interpreted as meaning any activity using the total cost approach, and in particular may be considered as comprising life-cost planning and life-cost analysis. Life-cost is the preferred term because it de-emphasizes the notion of a 'cradle-to-grave' life cycle that in reality is rarely applicable.

19.2 Life-cost planning

19.2.1 The purpose of cost planning

Cost planning is part of a wider cost management process that commences with the decision to build and concludes with the completion of design documentation. During this period the main objectives are:

1. The setting of cost targets, in the form of a budget estimate or feasibility study, as a framework for further investigation and as a basis for comparison.
2. The identification and analysis of cost-effective options.
3. The achievement of a balanced and logical distribution of available funds between the various parts of the project.
4. The control of costs to ensure that funding limits are not exceeded and target objectives are ultimately satisfied.
5. The frequent communication of cost expectations in a standard and comparable format.

The cost plan is one of the principal documents prepared during the initial stages of the cost management process. Costs, quantities and specification details are itemized by element (or sub-element) and summarized collectively. Measures of efficiency are calculated and used to assess the success of the developing design. The elemental approach aids the interpretation of performance by comparison of individual project attributes with similar attributes from different projects, and forms an extremely useful classification system. Life-cost plans differ from traditional capital cost plans only in the type of costs that are taken into account and how these can be expressed and interpreted.

Life-cost planning offers significant benefits on projects. Figure 19.1 illustrates some of the types of outcomes that can be expected, such as cost distributions and cash flow forecasts.

19.2.2 Types of costs

Life-costs can be divided into various categories that aggregate capital and operating expenses in different ways. The following categories are offered as a preferable division:

1. *Capital costs*. Capital costs comprise the initial acquisition of the land and building, and can include:
   - *Land cost*. The purchase cost of the land.
   - *Construction cost*. The cost of labour, material and plant involved in the creation of the building and other improvements to the land, including all supervision, profit, and rise and fall during the construction period.

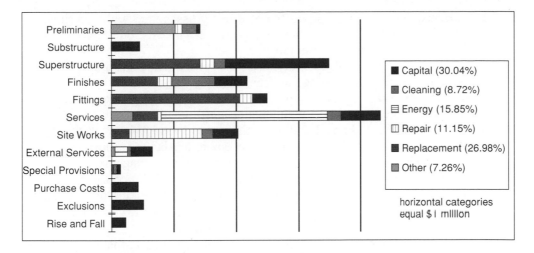

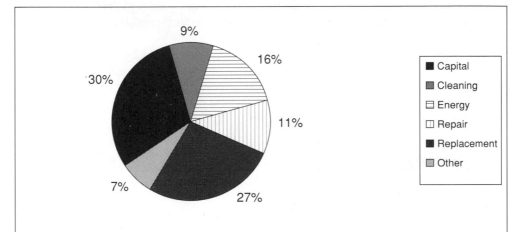

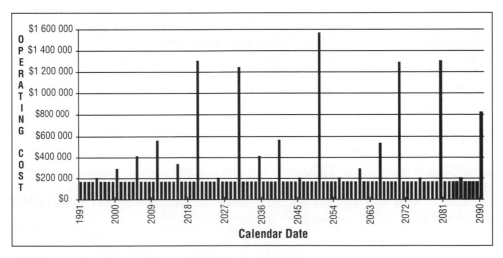

Figure 19.1 Examples of life-cost planning outcomes

- *Purchase cost*. Acquisition costs not directly associated with the finished product, including items such as stamp duty, legal costs, building fees, professional fees, and commissioning.
2. *Operating costs*. Operating costs comprise the subsequent expenditure required to service the land and building, and can include:
   - *Ownership cost*. Regular running costs such as cleaning, rates, electricity and gas charges, insurance, maintenance staffing, and security.
   - *Maintenance cost*. Annual and intermittent costs associated with the repair of the building including periodic replacement or planned renovation.
   - *Occupancy cost*. Costs of staffing, manufacturing, management, supplies, etc. that relate to the building's function, including denial-of-use costs.
   - *Selling cost*. Expenses associated with ultimate sale, including real estate agent commissions, stamp duty, transfer fees, etc. In some cases demolition or decommissioning costs may be considered.
3. *Finance costs*. Finance costs comprise expenditure relating to the interest component of loan repayments, establishment and account fees, holding charges and other liabilities associated directly with borrowed capital.

Life-costs are of a significant nature (Department of Quantity Surveying, 1988) and deserve consideration commensurate to their effect. Their absence from investment decision-making in the past has largely stemmed from the difficulties of applying these techniques and the tendency by professional advisers and building owners to ignore the impact of long-term expenditure.

19.3 Life-cost analysis

19.3.1 The purpose of cost analysis

The collection and interpretation of actual building costs as an input or feedback source for cost planning is the main function of traditional cost analysis. It includes distributing actual costs amongst the appropriate elements (or sub-elements) upon which cost plans are normally based, and has made possible a degree of control which was previously unknown. Composite rates can be averaged from real projects to provide a valuable guide to the estimation of new projects. This process applies equally well to life-costs as it does to capital costs alone.

Apart from providing feedback for future cost planning activities, analysis of life-costs can form an essential element of overall cost management by highlighting the ways in which potential cost savings in existing buildings might be achieved. For example, it might be better value to prematurely replace an expensive building component with a more efficient solution prior to the end of its useful life than to continue with a poor initial decision. Prudent control requires that actual and expected performance be constantly compared.

19.3.2 Monitoring and management

The monitoring and management of actual performance is known as life-cost analysis. While life-cost planning encompasses design cost control, life-cost analysis focuses on activities applicable to the construction and occupation of buildings. Together these techniques form the backbone of the cost management process and extend from the decision to build until the cost implications are no longer of concern.

Flanagan and Norman (1983) point out quite correctly that for a life-cost approach to be effective in reducing the running costs of existing buildings it is necessary that these running costs be continually monitored. Monitoring involves the recording of actual performance of a particular project in a form that facilitates subsequent life-cost planning and management activities. Such performance can be compared against the cost targets and frequency expectations given in the life-cost plan. Areas of cost overrun or poor durability can be explored with an aim to implement identified improvements.

19.3.3 Data collection

Data collection is a prominent activity in life-cost analysis. Data from different projects will reflect the specific nature of the building, its location and occupancy profile. When measured overall, no two buildings will have identical running costs, nor will the running costs for any specific building be the same from one year to another. Data collection is still a useful exercise, particularly for monitoring activities, but obtained results should be applied to other situations with caution.

19.4 Important considerations

19.4.1 Discounting

Discounting is a means by which equivalent value is determined. Costs and benefits which arise in different time periods must be brought to a common base so that a proper comparison can be made. This comparison concerns not just the timing of cash inflows and outflows for a given project, but their relationship across projects. The discounting process is relevant to other investment evaluation techniques, such as cost-benefit analysis, that employ a discounted cash flow methodology.

Discounting is merely a technique invented to help make judgements between investments that have different timing of costs expended or benefits received. Cash inflows that are received earlier and cash outflows that are incurred later are generally regarded as more preferable than the reverse. Time preference and capital productivity are two philosophies that can be used as the basis for the choice of discount rate. Discounting is based on the compound interest principle in reverse and can thus be described as a negative exponential. Over long time periods the effect, particularly at high rates of discount, can be such as to make future costs and benefits irrelevant. Figure 19.2 illustrates the effect of a 6% discount rate on the estimated operating cost cash flow for a project over a one hundred year life.

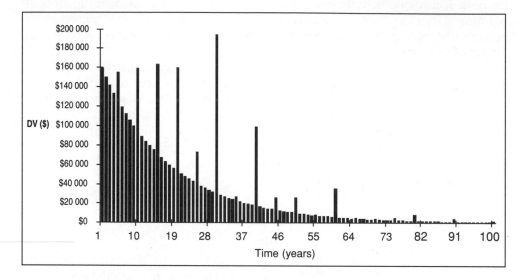

Figure 19.2 Example of the effect of discounting

Discounting is a commonly misunderstood process and is one of the principal reasons for the slow implementation of life-cost studies into normal practice in the construction industry. It is important to realize that discounting is an artificial mechanism invented so as to disadvantage future expenditure and income over the present for the purpose of equating the value of money in different time periods. The sum of all the discounted values does not represent real dollars, but is used solely for selection purposes. Therefore discounting can be correctly applied only when two or more alternatives are compared. Discounting includes for the cost of finance needed to fund the project, or the lost opportunity by not investing elsewhere, or both. Therefore interest is never included as a discrete item of expenditure, and to do so would be double counting.

While discounting has been described as socially unacceptable because it diminishes the impact of future cash flows and hence favours present rather than future generations, it nevertheless describes the basic tendency of all thinking people. Ask yourself whether you would prefer to receive $100 now or in one year's time. You would normally say that you would prefer the money now. Also ask yourself whether you would prefer to pay $100 now or in one year's time. You would normally say that you would prefer to pay later. This decision is based, at least in part, on the potential opportunity that money has when you have control over its use. Discounting is a mathematical representation of these otherwise subjective judgements, and enables us to assess complex patterns of expenditure and revenue with objectivity.

Discounting has relevance to life-cost studies where a comparison between a

number of alternative courses of action is involved. Results are expressed in terms
of discounted (present) value and the lowest value is the alternative that is the most
cost-effective. Table 19.1 illustrates an example of a typical comparison over a 25
year life using a discount rate of 2%. But discounting has no place in the measurement of a single design solution, as would occur in a life-cost plan for a proposed project, and in such cases it is recommended to present all costs in today's
terms (real value).

Discount rates based on the weighted cost of capital reflect the true time value
of money and do not distort the relationship between present and future events
unfairly. This approach is fundamental to comparisons over time because it
accounts for the costs of finance, either real or imputed. Underlying the exchange
rate for the time value of money is actually a zero discount rate. This signifies that
normally present and future generations are given equal weight.

Table 19.1 Worked example for hypothetical alternative comparison

	Alternative A	*Alternative B*
Data (real value)		
Capital cost	$15 000	$40 000
Cleaning cost	$100 pa	$100 pa
Energy cost	$4 000 pa	$2 500 pa
Maintenance cost	$500 pa	$500 pa
Intermittent maintenance cost (additional)	$500/5 yrs	$500/10 yrs
Equipment life	10 yrs	20 yrs
Replacement cost (less scrap value)	$11 000	$35 500
Comparison (discounted value)		
Capital cost	$15 000	$40 000
Cleaning cost	$1 952	$1 952
Energy cost	$78 094	$48 809
Maintenance cost	$10 891	$10 172
Replacement cost	$16 427	$23 890
Total life-cost	$122 364	$124 823
less residual value	($4 572)	($18 286)
less tax deductions (36% on operating)	($32 737)	($21 936)
less tax depreciation (10% PC on capital)	($8 358)	($14 030)
Comparative value (lowest is best)	$76 697	$70 571

19.4.2 Study period

The study period, or time horizon, is another important consideration. The study
period is commonly taken as equal to the economic life. This is the period of time
during which the project makes a positive contribution to the financial position of
the owners, both present and future. However, once the owner sells the property to
another person, the original owner is no longer concerned with its operating performance or the costs or revenues that it will subsequently incur or generate. From
the perspective of the original owner, the study period should logically be the
period over which the owner has a financial interest.

The determination of the study period for an investment project highlights the
dilemma between the often opposing objectives of investor interest and social

welfare. While it may be conceptually correct to optimize a project so that society achieves the maximum benefit possible, this strategy may also result in no detailed study ever being commissioned since the investor may be departing on a course that ultimately reduces the expected return. Issues of energy conservation and public sector investment suggest social welfare (economic life) is the most appropriate basis for the study period. Issues of profit maximization and private sector investment suggest investor interest (holding period) is preferable.

The choice of study period is, however, not as critical as might first be thought. In a comparison situation where discounting is employed, cash flows after 20 to 25 years will generally be rendered negligible to the outcome. Hence choosing a study period of 100 years, for example, is unnecessary. In a measurement situation where discounting is not employed, the study period can be seen as a continuum along which the property might be bought and sold. In this context the study period is best selected as the period of financial interest of the owner, as this will define the quantification process, but the period can normally be treated as a variable and 'what if?' decisions analysed based on a range of possible values.

Furthermore, government regulation at the time of planning approval application could force projects that are shown to be a cost burden for future generations to undergo redesign until their operation is within maximum allowances. This trend is already happening in various countries, and it may soon be routine to prepare a life-cost plan for all new projects, to undertake annual self-assessment of energy consumption and to be prepared for periodic external audits and the threat of financial penalties should actual performance exceed legislated limits.

19.4.3 Risk and uncertainty

The forecasting of future events is normally an integral part of the decision-making process. It also can be the subject of considerable uncertainty and therefore requires cognizance of the level of risk exposure. Discounting and life-cost studies are clearly reliant on appropriate forecasts of future events being made.

The decision-maker is faced with a fundamental choice. Either the uncertainty of the future can be ignored by dealing with only those matters that are known, or tools can be used to help make predictions. It is frequently accepted in the literature that it is preferable to plan even when the accuracy of the plan cannot be guaranteed, for otherwise decisions are made in isolation to the environment in which they will ultimately be judged.

A life-cost plan is an expectation of future design performance and hence forecasts events that are inherently uncertain. Even capital costs display a level of uncertainty, but when coupled with costs over a long period of time that themselves are a function of economic factors, obsolescence and operating performance the entire process may appear to be an exercise in futility. But the life-cost plan should be seen as a set of targets against which future events can be compared. While the quality of the original prediction is important, a far greater benefit can be realized if areas of poor performance or areas in which additional savings can be made are identified.

Comparative life-cost studies involve further aspects of uncertainty, notably the selection of the discount rate and the study period. Figure 19.3 illustrates a sensitivity analysis on the discount rate for a life-cost comparison between carpet and parquetry with and without tax considerations. But even these areas of uncertainty can be diminished to some extent since they always apply to a range of solutions and hence can have a corresponding effect on estimated costs and benefits across all alternatives.

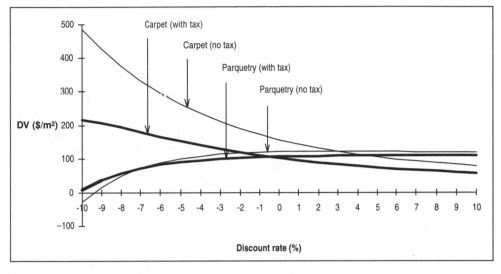

Figure 19.3 Example sensitivity analysis on the discount rate

Most forecasting techniques will inevitably fail to predict catastrophic events but instead will focus on a range of outcomes that may be reasonably concluded from history. This results in the identification of best, worst and most likely outcomes. If a decision is insensitive to this range it has a lower level of risk than if it were to easily change. Forecasting adds information to the decision-making process that is a vital part of proper analysis.

The adoption of risk analysis techniques enables the uncertainty of future events to be properly assessed. For example, if the discount rate applied to a life-cost study is likely to vary within a range of values, but at all values the project outcome is favourable, then what is at risk is the level of the benefit not the possibility of a loss. In this case uncertainty remains high but risk of financial loss is low. Investment decisions are not always so simple, and therefore a number of sophisticated risk analysis techniques are available to quantify the impact of various future scenarios. Risk analysis is a key tool in the assessment of value for money during pre-design.

19.5 Subjective considerations

Making decisions based on economic efficiency is useful, but overlooks the fact that options possess other attributes that cannot be readily translated into monetary models. Aesthetic quality, environmental impact, functional performance, flexibility and adaptability are examples of such attributes. Balanced decisions need to include these considerations either formally or informally.

A formal process is recommended. While judgement still plays a crucial role in any decision, a means of assessing overall value is an advantage in that it can make the rationale for design choices more objective and defendable. The aim is to be able to identify value for money, which in this context is defined as the ratio of functional performance to the cost of delivery.

19.5.1 Value management

Value management is a form of multi-criteria analysis commonly employed in the construction industry to assess and optimize value. It is a structured technique that works in a multi-disciplined environment. It can be used at various stages in the development process, assisting with pre-design, design and post-design decisions. Brainstorming and group dynamics are important ingredients in the successful pursuit of value for money. It can apply to any project, system, process or component that has a reasonable possibility for value improvement.

Value management draws on a methodological framework comprising information, speculative, analytical and proposal phases (Dell'Isola and Kirk, 1995a). These phases are intended to be completed sequentially. Each phase has a particular purpose. For example, the information phase is designed to understand the primary and secondary functions required and therefore properly understand the problem. The speculative phase is about idea generation. The analytical phase is about objective evaluation. The proposal phase is about making a recommendation, including implementation and follow-up actions.

Value management is sometimes criticized as being nothing more than organized commonsense. But its strength lies in its methodical approach and the open-minded investigation processes performed in a group environment (known as a VM workshop). There is sufficient evidence to conclude that use of value management can deliver advantages greater than its own cost.

In its traditional form, value management uses a matrix evaluation system to present options as rank scores. This is convenient in that all aspects of a decision can be identified and assessed in a common unit. Alternatives are presented in rows and scored (usually on a scale of 0–5) against a number of performance criteria presented in columns (usually weighted on a scale of 1–10). The multiplication of score and weight is accumulated across all criteria for each alternative to create a value index. Accuracy for the scores and weights is often best treated as whole numbers rather than decimals. The highest score indicates the best alternative.

19.5.2 Value for money index

The disadvantage of the traditional approach is that it converts criteria well suited to monetary evaluation into a less tangible format, and often disguises the importance of financial criteria. It therefore can be both misleading and manipulative in situations where the criteria and weightings are not properly conceived.

A better approach is to completely separate objective and subjective criteria until the final step. In this case, a weighted evaluation matrix is used to assess rank scores for all criteria that cannot be objectively measured in terms of life-cost. Issues like durability, maintenance, ease of construction, ease of cleaning, etc. are therefore excluded from the matrix. A value score is created for each alternative and represents functional performance achievement. The next step is to calculate the discounted present value of each alternative over the nominated study period. The derived life-cost is inclusive of capital and operating issues, and financial costs are absorbed into the discounting process. The final step is to divide value score by life-cost. The result is a value for money index – the higher the index the better.

The final decision is based on this index, as it takes into account all known considerations. The process assumes that functional performance and life-cost are both relevant to a decision. In cases where cost is all that matters, decisions can be made purely on this attribute. In cases where cost is irrelevant, decisions can be made purely on functional criteria. But most development decisions seek value for money, which can be translated as maximum functional performance for minimum cost. Table 19.2 illustrates the process.

19.6 Conclusion

Past analyses of design solutions for building projects have concentrated on initial capital costs, often to the extent where the effects of subsequent operating costs are completely ignored. However, even in cases where a wider view of cost has been adopted, the discounting process has commonly disadvantaged future expenditure so heavily as to make performance after the short term irrelevant to the outcome, resulting in projects which display low capital and high operating costs to be given favour. Thus design solutions that aim to avoid repetitive maintenance, reduce waste, save non-renewable energy resources or protect the environment through selection of better quality materials and systems, usually having a higher capital cost, are often rejected on the basis of the discounting process (Langston, 1991a).

Life-cost studies comprise both measurement and comparison activities. Measurement relates to the total cost of a particular design solution and involves the establishment of expected performance targets and their monitoring over time. Comparison relates to a number of alternative design solutions and in addition to total cost (and in many cases differential revenue) must consider the relative timing of receipts and payments. Most of the literature concentrates on the latter activity and hence gives the impression that discounting plays an absolute role in life-cost studies. But this outcome is disputed, and it is recommended that measurement activities are important and that their presentation in terms of real value is more meaningful than the use of discounted value.

Table 19.2. Enhanced Value Management Decision Process

PRIMARY (BASIC) FUNCTION: External paving for uncovered site paths							

PERFORMANCE CRITERIA

Criteria A = safety	Criteria weighting (0–10)	= 10
Criteria B = stability		= 8
Criteria C = appearance		= 5
Criteria D = water shedding ability		= 4

SHORT-LISTED ALTERNATIVES: weighting:	A 10	B 8	C 5	D 4	VALUE SCORE (rank score)	LIFE-COST (discounted $/m^2)	VALUE FOR MONEY (index)
brick pavers on sand base	3 / 30	4 / 32	5 / 25	5 / 20	107	$67.37	1.59
brick pavers on concrete slab	4 / 40	5 / 40	5 / 25	5 / 20	125	$76.18	1.64
concrete pavers on sand base	3 / 30	4 / 32	4 / 20	5 / 20	102	$62.21	1.64
concrete pavers on concrete slab	4 / 40	5 / 40	4 / 20	5 / 20	120	$71.02	1.69
concrete slab with wood float finish	5 / 50	5 / 40	3 / 15	5 / 20	125	$49.41	2.53
concrete slab with broom finish	5 / 50	5 / 40	4 / 20	5 / 20	130	$49.91	2.60
concrete slab with aggregate finish	5 / 50	5 / 40	5 / 25	5 / 20	**135**	$67.39	2.00
concrete slab with quarry tile finish	5 / 50	4 / 32	5 / 25	5 / 20	127	$113.79	1.12
concrete slab with cement topping finish	5 / 50	4 / 32	5 / 25	5 / 20	127	$65.89	1.93
natural stone paving	3 / 30	4 / 32	5 / 25	5 / 20	107	$170.91	0.63
precast concrete blocks	3 / 30	4 / 32	4 / 20	5 / 20	102	$224.22	0.45
pine bark mulching	2 / 20	1 / 8	5 / 25	3 / 12	65	$28.75	2.26
loose aggregate	1 / 10	1 / 8	5 / 25	3 / 12	55	**$22.73**	2.42
asphaltic concrete	5 / 50	4 / 32	2 / 10	5 / 20	112	$33.56	**3.34**

FINAL SELECTION: Asphaltic concrete (value for money index = 3.34).

Notes:
Concrete slab with aggregate finish has the highest functional performance (135).
Loose aggregate has the lowest cost ($22.73) and is known as the value standard.

References and bibliography

Ashworth, A. (1999). *Cost Studies of Buildings* (3rd edition). Longman.
Boardman, A.E. (1996). *Cost-Benefit Analysis: Concepts and Practice*. Prentice Hall.
Bull, J.W. (1992). *Life Cycle Costing for Construction*. Thomson Science and Professional.
Caimbrone, D.F. (1997). *Environmental Life Cycle Analysis*. Lewis Publishers.
Daly, H.E. (1997). *Beyond Growth: The Economics of Sustainable Development*. Beacon Printing.
Dell'Isola, A.J. and Kirk, S.J. (1995a). *Life Cycle Costing for Design Professionals* (2nd edition). McGraw-Hill.
Dell'Isola, A.J. and Kirk, S.J. (1995b). *Life Cycle Cost Data* (2nd edition). McGraw-Hill.
Department of Quantity Surveying (1988). *Home Life-Costs*. University of Technology, Sydney.
Diesendorf, M. and Hamilton, C. (1997). *Human Ecology, Human Economy*. Allen and Unwin.
Fabrycky, W.J. and Blanchard, B.S. (1991). *Life-Cycle Cost and Economic Analysis*. Prentice Hall.
Field, B.C. (1997). *Environmental Economics: An Introduction* (2nd edition). McGraw-Hill.
Flanagan, R. and Norman, G. (1983). *Life Cycle Costing for Construction*. Surveyors Publications.
Flanagan, R. and Norman, G. (1993). *Risk Management and Construction*. Blackwell Science.
Flanagan, R. and Tate, B. (1997). *Cost Control in Building Design: An Interactive Learning Tool*. Blackwell Science.
Flanagan, R., Norman, G., Meadows, J. and Robinson, G. (1989). *Life Cycle Costing: Theory and Practice*. BSP Professional Books.
Goodstein, E.S. (1998). *Economics and the Environment* (2nd edition). Prentice Hall.
Hanley, N., Shogren, J.F. and White, B. (1996). *Environmental Economics in Theory and Practice*. Oxford University Press.
Langston, C. (1991a). *The Measurement of Life-Costs*. NSW Department of Public Works.
Langston, C. (1991b). *Guidelines for Life-Cost Planning and Analysis of Buildings*. NSW Department of Public Works.
Langston, C.A. (1993). 'Comparative Life-Cost Studies'. *The Building Economist*, March, pp. 11–13.
Langston, C. (1994). 'The Determination of Equivalent Value in Life-Cost Studies: An Intergenerational Approach'. PhD Dissertation, University of Technology, Sydney.
Langston, C. (1996). 'Life-Cost Studies'. In BDP *Environment Design Guide* (GEN 10), Royal Australian Institute of Architects.
Pearce, D.W., Markandya, A. and Barbier, E.B. (1989). *Blueprint for a Green Economy*. Earthscan Publications.
Price, C. (1993). *Time, Discounting and Value*. Blackwell Publishers.
Seeley, I.H. (1997). *Building Economics*. Macmillan Press.
Thomas, R. (1999). *Environmental Design: An Introduction for Architects and Engineers* (2nd edition). E. & F.N. Spon.
van Pelt, M.J.F. (1993). *Ecological Sustainability and Project Appraisal*. Avebury.

Determination of the discount rate

20.1 Introduction

When evaluating construction projects in terms of their environmental impact, due consideration needs to be given to all the costs and benefits that flow from the decision over the life of the project. This immediately raises the question of how to equate costs and benefits that occur in different time periods. Discounting is a technique that has been developed to adjust for the effects of time by recognizing a preference for events that occur in the near future over events that occur in the far future. The rate of discount used is critical to the evaluation process and the validity of the outcome. There is a clear connection between the environmental crisis, economic development, the need for such development to be sustainable and the role that discounting plays in the determination of equivalent value.

Past analyses of design solutions for construction projects have concentrated on initial capital costs, often to the extent where the effects of subsequent operating costs are completely ignored. However, even in cases where a wider view of cost has been adopted, the discounting process has commonly disadvantaged future expenditure so heavily as to make performance after the short term irrelevant to the outcome, resulting in projects which display low capital and high operating costs to be given favour. Thus design solutions that aim to avoid repetitive maintenance, reduce waste, save non-renewable energy resources or protect the environment through selection of better quality materials and systems, usually having a higher capital cost, are often rejected on the basis of the discounting process (Langston, 1991).

The factors influencing the choice of discount rate are of concern because sustainable development is possible only if the equivalent value of future performance can be appropriately assessed. Although economic progress and environmental conservation are not mutually exclusive, there is normally a trade-off and a balance therefore needs to be struck. Bias towards one or the other will directly affect the quality of life of future generations. Discounting must facilitate the equitable assessment of both present and future events. Doubt exists over whether this is currently the case.

There has been substantial research into the process of discounting, much within the last few decades, which suggests that it is either a very complex subject or a very confused one (perhaps both). Angelsen (1991, p. 12) writes:

> A discussion of the arguments for discounting raises several problems. The literature is anything but clear, and there exists little consensus on the subject. Different theories lead to different conclusions, and the positions are hard to compare as the assumptions and approaches differ considerably. According to Dixon and Meister (1986, p. 41) discounting is 'one of the most misunderstood concepts in economic analysis'. Two decades earlier, Baumol (1968, p. 788) similarly noted that 'few topics in our discipline rival the social rate of discount as a subject exhibiting simultaneously a very considerable degree of knowledge and a very substantial level of ignorance'. In the theoretical literature, complex models are developed to find the appropriate rate, whereas in practical situations one finds rather pragmatic judgements.

Yet on the whole economists have acted as if there was no controversy and no broad constituency of concerns; they have routinely and mechanically applied the discounting technique in arenas of human activity very far removed from the financial markets in which they have their origin. There is great institutional convenience in the uniform adoption of a standard form of appraisal; it seems to offer consistency, even if only in the form of consistent error. Compromises have been sought that modify discounting incrementally so that its long-term effects are less severe. But most agree that discounting is so firmly entrenched that it cannot readily be displaced, even if such action was considered appropriate (Price, 1993).

20.2 The rationale for discounting

Discounting is defined as the means by which equivalent value is determined. Costs and benefits that arise in different time periods must be brought to a common base so that a proper comparison can be made. This comparison concerns not just the timing of cash inflows and outflows for a given project, but the relationship of cash flows across projects. Therefore discounting is a decision-aiding technique that enables alternative investments to be evaluated and ranked.

Discounting processes result in the calculation of 'equivalent' values that are useful in comparing alternative solutions involving cash flows that span over more than one year. Discounted value is an 'abstract' measure that negates the need to explicitly consider finance costs while simultaneously allowing for the timing of cash flows. Ashworth (1999) warns that we need to remind ourselves consistently that we are comparing the worth of alternative systems and not their costs. Robinson (1986, pp. 18–19) confirms that in the context of life-costs:

> The problem areas generally arise as such because potential users of the technique are misled by the idea that the life cycle cost outcome must stand by itself in absolute terms. A little thought will lead to the conclusion that this is not possible. The life

cycle cost outcome is only useful as a comparative figure for the purpose of ranking one solution over another. The same thing can be said of all the economic evaluation techniques which employ discounted cash flow analysis.

The necessity to take account of the time value of money when comparing future costs and benefits is clear, but the mechanism employed to achieve this is not. Discounting is advocated generally in the literature as being the appropriate conversion tool, and indeed has widespread acceptance in appraising investment projects in practice. Nevertheless, Flanagan (1985, p. 13) states that 'many consultants and clients, whilst understanding the concept, are not totally comfortable with the use of discounting as a process'. Dawson (1986, p. 14) confirms this view and observes that 'there still seems to be an aura of mystique about discounting'. There may be some validity to these perceptions.

One argument against the discounting process is that it renders future costs and benefits insignificant when a lengthy time horizon is involved. Öfverholm (1987, p. 613) states that 'the traditional discounting method reduces future costs to such an extent that they lose almost all their impact on decisions whether to invest in more durable designs and materials [. . . and] has given rise to doubt as to the value of the method'. Urien (1975) states that for expenses in the very far future, the classical formulae for discounting do not have much meaning, since they 'dilute', to a considerable extent, the impression made by the distant expenditures. But Stone (1974, p. 73) conversely believes 'this is a convenient feature of the technique since the relative weight given to consequences tends to decline with their degree of certainty [and] thus the greatest weight tends to be given to the costs of development about which there is the greatest certainty'.

Others have criticized discounting on social grounds. Pearce and Markandya (1987), for example, do not consider discounting to be socially justified. They conclude that as far as the balance between the present and future generations is concerned, discounting gives more importance to the former. Ramsey (1931) describes discounting as immoral and Pigou (1932) suggests that the technique was invented to overcome the defective telescopic faculty of the human mind. Discounting may lead to decisions being made that are inadvertently biased against future generations.

Stone (1983, p. 201) states that 'future costs are worth less than their face value because they can be secured by investing a smaller sum now, which with interest will grow to the sum required at the date in the future when it is required'. Bromilow (1985) continues this theme by explaining that the effect of discounting is to estimate the value of a fund which, if provided at the inception of the project, would finance not only the initial capital investment but have enough left over, together with the interest it would earn when reinvested, to pay for the recurrent expenses over the life of the building. Of course, the reality of having such a fund available from the outset is incredible. Discounting must be seen as a theoretical value adjustment only.

The choice of discount rate, if not made with due consideration to investment return, inflation and taxation, can lead to significant misinterpretation of decisions. Use of high interest rates based on project selection criteria has had the effect of minimizing the benefit of increased operating efficiency and has fuelled speculation that discounting may not be totally appropriate. Large discount rates are a fallacy in today's economic climate, and perhaps always were so. Discounting with pure interest rates or borrowing rates is invalid unless the costs being discounted are first inflated (Goodacre, 1978). Ignoring income tax liability on investment return for non-government organizations is unrealistic, even though some companies may, from time to time, show a small net profit or even a loss at the end of a financial year. It can be argued that the choice of discount rate is one of the more critical variables in the analysis (Ashworth, 1999).

Stone (1975, p. 70) acknowledges that discounting has drawbacks, but suggests that 'the value of the technique lies in the rigour introduced into the evaluation, the need to make all assumptions explicit and the separation of real and chance effects'. Discounting is one method for objectively depicting the time value of money, but there is still no universal measure of comparison between costs incurred now and those incurred in the future (Tucker, 1987). Furthermore, Bowen (1974) concludes that discounted values are practically meaningless unless compared to that of an alternative solution.

Nevertheless, most people would prefer to receive revenue sooner rather than later and would prefer to postpone expenditure for as long as possible. This is sometimes referred to in the literature as the personal rate of time preference or pure time preference. In order to increase the investment return, initial revenue must be maximized and initial expenditure minimized. Discounting future costs with a positive discount rate has the effect of favouring initial receipts and payments over those that occur subsequently. Assets that possess low initial costs and defer operating costs into the future lead to higher returns for the investor.

While there is a rationale for discounting as a process, the selection of the rate of discount is the subject of much controversy. Pearce et al. (1989) identify two distinct philosophies for discount rate selection. These are time preference and capital productivity.

20.3 Time preference

The philosophy of time preference is based on the assertion that people are impatient and seek control over money, assets and other resources. Individuals prefer the present to the future, even where monetary gain is not involved. As society is no more than the sum of individuals, therefore society also prefers the present to the future. This temporal preference translates into a positive discount rate (Markandya and Pearce, 1988). Present consumption is therefore preferred to past consumption. Further analysis of the time preference philosophy reveals that there are at least three distinct supporting reasons. These are:

1. *Pure impatience*. Impatience (or myopia) is an undeniable human trait that results in present events being more desirable than those occurring in the future.
2. *Risk and uncertainty*. A benefit or cost is valued less the more uncertain its occurrence. This can result from uncertainty about the presence of the individual at some future date (risk of death), uncertainty about the preferences of the

individual even when his or her existence can be regarded as certain, and uncertainty about the availability of the benefit or existence of the cost.

3. *Intergenerational equity*. If future generations are confidently expected to be 'better off', then the value of future events can be given less weight compared to the value of present events.

These factors may combine or be subsumed in some way to determine what is commonly identified in the literature as the social time preference (STP) rate of discount. Although impatience may well exist and be relevant to the calculation of equivalent value, it is difficult to quantify other than to assume that it can be reasonably modelled by use of normal investment return.

20.4 Capital productivity

The philosophy of capital productivity is based on the theoretical earning capacity of money over time. While this phenomenon is real, there is little consensus to the method of calculation of the discount rate. Most opinions, however, can be reasonably aligned with one of the following assertions:

1. The discount rate represents the minimum acceptable rate of return for the project.
2. The discount rate represents the opportunity cost sacrificed by not investing in an alternate project.
3. The discount rate represents the weighted cost of capital applicable to the financing of the project.

Peirson and Bird (1981, p. 130) state that the 'project's net present value [shall be calculated] using the required rate of return as the appropriate discount rate'. If the net present value is positive, the project will generate at least the required rate and is thus acceptable. Determination of the minimum acceptable rate or return (MARR) may be based on a variety of factors, but will typically reflect the interest cost of borrowed funds (preferably after tax) plus an allowance for profit and risk (Haviland, 1977). Flanagan (1984) further suggests that it would be feasible to discount 'more risky' projects more heavily in order to allow for risk.

Alternatively the discount rate can be based on opportunity cost, defined as meaning the real rate of return on the best available use of funds to be devoted to the proposed project (Ashworth, 1999). Ruegg and Petersen (1987) add that the best available use of funds should represent a similar level of risk. Public projects often use discount rates based on the social opportunity cost of capital (SOC), which is interpreted as displaced private spending. The SOC is the marginal productivity of available capital and has been identified as the best prospect for a practical approach to discounting for project analysis in the Australian public sector (Treasury Department, 1978).

While the minimum acceptable rate of return and opportunity cost approaches concerning the choice of discount rate are not greatly dissimilar, the third assertion distinguishes itself by not specifically involving profit or risk. The latter attempts to reflect the true time value of money only based on explicit or implicit investment return rather than desirable levels of profitability.

Investors must either borrow money to finance the project or sacrifice interest earned on their own money (Stone, 1973). Use of investment return for discounting implies that interest on present equity for the full cost of the investment is being foregone, interest on borrowed funds for the full cost of the investment is being charged, or more likely a combination of both equity and borrowed funds is involved. Ideally the discount rate should be based on a weighted mix of the interest rate sacrificed by use of equity and the interest rate payable by use of borrowed funds, representing the expected financing arrangements for the project (Dell'Isola and Kirk, 1995a).

Furthermore, the selection of an appropriate discount rate to be used in the calculations will largely depend upon the financial status of each individual investor (Ashworth, 1999). The ratio of debt to equity financing for a company has been used to determine the weighted cost of capital. Whichever capital productivity approach is employed for determining the discount rate, it can be proven that they:

- negate the need to include interest payable on borrowed funds in the cash flow, as this is reflected in the establishment of the discount rate (Dell'Isola and Kirk, 1995a).
- favour expenditure in the far future over expenditure in the near future by enabling the saved equity to theoretically earn interest.
- are not affected by whether particular cash flows are financed from equity or borrowings provided that the overall mix of capital sources is appropriate.

20.5 Further considerations

A choice between time preference and capital productivity, or some combination of both, is therefore necessary before a discount rate can be selected. Allied to these philosophies, however, are a number of further considerations. These can be summarized as comprising:

1. *Risk.* Some projects contain a higher level of risk than others and so may have a higher discount rate to account for the uncertainty of future events. This is clearly an aspect of time preference and may be considered to be included in the opportunity cost of capital.
2. *Profit.* High profit expectations will similarly result in high discount rates so that only projects that satisfy these expectations are selected.
3. *Taxation.* The incorporation of taxation matters, where applicable, will lower discount rates yet will increase the individual cash flows that occur in the future due to income tax deductions and depreciation allowances.
4. *Environmental issues.* The potential for depletion, catastrophic loss or irreversibility of non-renewable natural resources, in particular, might encourage higher discount rates to be selected so as to constrain development. Conversely, low discount rates might favour projects that have efficient operating demands and hence also contribute to the conservation of resources.

Generally it is expected that a discount rate based on time preference would be lower that one based on capital productivity. Willis (1980) argues that the latter takes adequate account of the former, and recommends that opportunity cost be adopted as the appropriate solution. But if the discount rate is based on time preference then inclusion of actual interest costs into the cash flow must occur, and thus while the discount rate may be lower the net present value may not necessarily be higher.

The composition of the discount rate is a controversial issue. There are a number of components that may or may not be incorporated into the discount rate. In some cases this may depend on the technique being used and the purpose to which it is being put.

Under the capital productivity philosophy to discounting, possible components of the discount rate may include:

1. Investment return (comprising a weighted mix of interest rates for equity use and borrowed monies).
2. General inflation.
3. Taxation.
4. Profit expectations.
5. Risk.

Under the time preference philosophy to discounting, possible components of the discount rate may include:

1. Pure impatience.
2. Risk and uncertainty.
3. Intergenerational equity.

Most literature suggests that the discount rate is normally based on capital productivity. Using this approach the discount rate reflects, at least in part, the interest payable on borrowed money to finance the investment and/or the interest lost through use of accumulated equity. A weighted combination of both interest rates is therefore involved. The interest rates may be real or nominal, before or after tax and may additionally include or exclude profit expectations and risk contingencies.

Matters of a non-financial nature are not well represented by the compound interest approach. It can be argued, for example, that a holiday in Hawaii is preferred now to one later in the year (all other factors being equal) yet this preference appears unrelated to the growth in reinvested income. Similarly, risk levels will influence the expected estimates for individual cash flows, perhaps by raising all costs by 10% and lowering all benefits by 10%, but this outcome is quite different to that resulting from a 10% increase in the discount rate.

On the other hand, financial matters can be realistically modelled using a compound interest approach. Interest payable for monies borrowed or interest lost due to equity usage both result in a burden that is proportional to time. The source of financing to be used for a project is relatively unimportant.

Interest can be incorporated into the discount rate or dealt with explicitly as a cash inflow or outflow. It may be correct to ignore discounting altogether provided an interest component is included in the annual cash flows as an expense for every year of the project. However the use of a discount rate reflecting the weighted mix of equity and borrowing gives an identical result. There is no doubt that the inclusion of interest into both the cash flows and the discount rate is double counting and thus incorrect.

A discount rate limited to the real weighted cost of capital reflects the true time value of money. The profitability of a project and its associated risk, of vital concern to an investor, can be independently assessed. Profit and risk expectations do not behave in an exponential manner, which is a further reason why they should be eliminated from the discount rate. Given that the entire discounting process is based upon the compound interest formula, it is reasonable when deciding on the composition of the discount rate to include only those components that are compatible.

Since interest is intended in part to offset the devaluing effects of inflation and is expressed in 'future cost' terms, the interest cash flows would need to be reduced by the expected inflation rate. Furthermore, testing the sensitivity of assumptions in isolation or combination will take on overwhelming proportions. For all these reasons it is desirable to allow the discount rate to incorporate interest transactions.

If interest was to be treated as a distinct cash flow then the entire discounting process may well be redundant. Therefore it can be concluded that a primary purpose of discounting is to allow for interest lost or payable. Discounting deals with the preference for early receipt of benefits or late payment of costs through the interest principle.

Investment return for use in the discount rate must be a mix of interest rates for both lost equity and borrowed funds utilized in the project. One of the issues that needs to be resolved, therefore, is the ratio of equity to borrowings and this ratio must be applicable to all the cash flows that are expected to occur over the selected study period.

20.6 Discount rate determination

The literature frequently describes the composition of the discount rate in terms of investment return, inflation and taxation. Formulae are also presented that can calculate the discount rate to any number of decimal places. Yet in practice it is common to assess the discount rate rather than to calculate it. This assessment is often based on personal judgement, intuition and previous experience. It is therefore difficult under these circumstances to know exactly what the discount rate is incorporating and this may have contributed over time to the general uneasiness that practitioners have about the discounting process.

Government authorities increasingly set discount rates for economic appraisal based on Treasury advice. Three rates are used, called 'test' discount rates, and are based on likely, pessimistic and optimistic scenarios. For example, the advocated discount rate may be 7% but a risk analysis is required using both 4% and 10% to highlight the sensitivity of the investment. If the project is viable at all discount rates then a degree of confidence in the proposal is achieved.

The discounting process is better served by removing those factors that are not well represented by the compound interest approach. The discount rate will therefore be based on investment return, inflation and perhaps taxation. Profit expectations and risk contingencies are dealt with separately. Such a strategy, apart from being conceptually correct, will demystify much of the theory surrounding the discounting technique and will lead to greater understanding and confidence.

It is concluded that the discount rate should be equal to the after-tax weighted average of equity and borrowed capital applicable to the project being appraised. The individual cash flows should include income tax deductions, depreciation and other taxation concessions or liabilities but must exclude all interest received or payable. Individual cash flows should preferably be expressed in present-day terms (real terms) and if so the discount rate must be exclusive of inflation. Profit and risk contingencies are assessed external to the discounting process.

The weighted cost of capital, after inflation and taxation have been removed, is unlikely to be high. Under certain conditions the resultant discount rate may even be negative. Although it may seem incongruous that any discount rate can be less than zero, in reality this situation highlights that the after-tax investment return is offset by inflation to the point where money is devaluing over time despite its productive use. Future cash flows therefore increase in comparative value.

It is resolved that the discounting process is merely a technique available for comparative purposes that disadvantages future costs and benefits relative to present values. Discounting is based on the compound interest principle in reverse and thus can be described as a negative exponential. Over long time periods the effect, particularly at high rates of discount, can be such as to make future costs and benefits irrelevant.

After critical analysis of the literature, the following conclusions about the factors that influence the choice of discount rate can be drawn:

1. The discount rate should be based on forecasts of investment return, inflation and taxation appropriate to the selected study period.
2. Interest payable on borrowings or interest lost on use of equity should not be explicitly included in the cash flow forecasts but dealt with by the discount rate.
3. Discount rates should be real (i.e. inflation should be removed) and the cash flow forecasts should be expressed in present-day terms.
4. Taxation deductions, depreciation and liabilities should be explicitly included in the cash flow forecasts and the discount rate should be net after tax.
5. Profit should not be built into the discount rate but its adequacy should be judged by the difference between the discount rate and the internal rate of return.
6. Risk should not be built into the discount rate but should be assessed separately using one or more specialist risk analysis techniques.
7. The discount rate should be calculated as the after-tax weighted average of equity and borrowed capital applicable to the individual projects being appraised.

The discount rate is project-related, not purely market-related as is generally understood in the literature, and therefore may differ between competing investment opportunities. Furthermore, the use of artificially high rates of discount unreasonably distorts the equitable balance of initial and subsequent cash flows. While a capital productivity approach is advocated in general, time preference issues are relevant: impatience can be modelled by the theoretical interest lost on equity, risk and uncertainty are grouped with other risk considerations and analysed as a separate process and intergenerational equity appears conceptually valid.

In addition to the matters discussed above, research into discounting by Langston (1994) has shown that consideration of opportunity cost alone is an over-simplification. Changes in the affordability of goods and services and the purchasing power of available income affect the value of future cash flows, and hence should also be incorporated into the discount rate. This approach involves a combination of time preference and capital productivity and allows for intergenerational variations to be modelled.

The determination of the discount rate can be based on the following formula:

$$\text{Discount rate (d)} = \frac{(1 + (i_1 p_1 + i_2 p_2)(1 - t)) - 1}{(1 + e)(1 + r)}$$

where:
i_1 = interest rate for equity p.a. (%) divided by 100
i_2 = interest rate for borrowing p.a. (%) divided by 100
p_1 = proportion of equity (%) divided by 100
p_2 = $1 - p_1$
t = taxation rate p.a. (%) divided by 100
e = escalation rate p.a. (%) divided by 100
r = affordability rate p.a. (%) divided by 100

20.7 Conclusion

The discount rate has been used in practice to account for a range of factors. It is argued that the loading of the discount rate is not an appropriate strategy. The outcome of this action is to distort the emphasis of initial costs and benefits relative to those that occur later, often resulting in the rejection of alternatives that avoid repetitive maintenance, reduce waste, save non-renewable energy resources, protect the environment or otherwise exhibit efficient operating performance.

Furthermore it is commonly stated in the literature that the discounting process discriminates against unborn future generations. Yet where the discount rate is based on the real weighted cost of capital, as advocated in this paper, no discrimination is implied since the rate is merely allowing for the cost of borrowing money or losing interest on equity which otherwise would need to be considered as an explicit cash flow.

Discounting is a time-honoured institution. While modifications may be permissible on an incremental basis, wholesale rejection of the process is both unlikely and unwarranted. It is clear that some form of adjustment for time is necessary, but what is questioned is the manner in which this adjustment is determined and applied.

Sustainable development has introduced a new dimension to the economic appraisal of investments. No longer can the environment be treated as an inexhaustible supply of resources nor a dumping ground for waste products. Environmental impacts must be integrated into investment decisions by the recognition that they have value and that this value does not necessarily change over time in the same manner as reinvestible capital. The commonly expressed argument that future prosperity is achieved by the accumulation of capital wealth alone is disputed.

References and bibliography

Angelsen, A. (1991). *Cost-Benefit Analysis, Discounting and the Environmental Critique: Overloading of the Discount Rate?* Chr. Michelsen Institute, Department of Social Science and Development, Norway.

Ashworth, A. (1999). *Cost Studies of Buildings* (3rd edition). Longman.

Baumol, W.J. (1968). 'On the Social Discount Rate'. *American Economic Review*, vol. 58, pp. 788–802.

Bowen, B. (1974). 'Design Evaluation Techniques'. *Progressive Architecture*, May, pp. 116–17.

Bromilow, F.J. (1985). 'Life Cycle Costing of Buildings and Services'. In proceedings of AIRAH TECH-85 Conference, Melbourne.

Caimbrone, D.F. (1997). *Environmental Life Cycle Analysis*. Lewis Publishers.

Dawson, B. (1986). 'Life Cycle Costing: Case Studies from a Contractor's Perspective'. In proceedings of Building Science Forum of Australia (NSW Division) Seminar, Sydney, pp. 13–18.

Dell'Isola, A.J. and Kirk, S.J. (1995a). *Life Cycle Costing for Design Professionals* (2nd edition). McGraw-Hill.

Dell'Isola, A.J. and Kirk, S.J. (1995b). *Life Cycle Cost Data* (2nd edition). McGraw-Hill.

Department of Housing and Construction (1979). *Life Cycle Costing* (Technical Information TI 140 AE). Canberra.

Dixon, J.A. and Meister, A.D. (1986). 'Time Horizons, Discounting and Computational Aides'. In *Economic Valuation Techniques for the Environment* (J.A. Dixon and M.M. Hufschmidt, eds.), Johns Hopkins University Press.

Flanagan, R. (1984). 'Life Cycle Costing: A Means for Evaluating Quality'. In *Quality and Profit in Building Design* (P.S. Brandon and J.A. Powell, eds.), E. & F.N. Spon, pp. 195–207.

Flanagan, R. (1985). 'Life Cycle Costing: What has Happened in the United Kingdom'. In proceedings of International Comparison of Project Management and Performance Conference, Melbourne.

Flanagan, R. and Tate, B. (1997). *Cost Control in Building Design: An Interactive Learning Tool*. Blackwell Science.

General Services Administration (1975). *Life-Cycle Costing in the Public Buildings Service*. Booz, Allen and Hamilton Inc.

Goodacre, P. (1978). 'Reappraisal of Cost-In-Use Calculations'. *Building Technology and Management*, vol. 16, no. 5, pp. 6–7.

Haviland, D.S. (1977). *Life Cycle Cost Analysis: A Guide for Architects*. The American Institute of Architects.

Langston, C.A. (1991). *The Measurement of Life-Costs*. Public Works Department of New South Wales, Sydney.

Langston, C.A. (1994). 'The Determination of Equivalent Value in Life-Cost Studies: An Intergenerational Approach'. PhD Dissertation, University of Technology, Sydney.

Markandya, A. and Pearce, D.W. (1988). 'Environmental Considerations and the Choice of Discount Rate in Developing Countries' (Environmental Department Working Paper No. 3), World Bank, Washington.

National Public Works Conference (1988). *Life Cycle Costing*, Canberra.

Öfverholm, I. (1987). 'Viewpoints on Steering Models for Life-Cycle Costs of Buildings'. In *Building Cost Modelling and Computers* (P.S. Brandon, ed.), E. & F.N. Spon, pp. 611–17.

Pearce, D.W. and Markandya, A. (1987). 'The Benefits of Environmental Policy' (Discussion Paper). University College, Department of Economics.

Pearce, D.W., Markandya, A. and Barbier, E.B. (1989). *Blueprint for a Green Economy*. Earthscan Publications.

Peirson, G. and Bird, R. (1981). *Business Finance* (3rd edition). McGraw-Hill.

Pigou, A.C. (1932). *The Economics of Welfare*. Macmillan.

Price, C. (1993). *Time, Discounting and Value*. Blackwell.

Przybylski, E.A. (1982). 'Effect of Inflation and Rate of Interest on Cost-In-Use Calculations'. In *Building Cost Techniques: New Directions* (P.S. Brandon, ed.), E. & F.N. Spon, pp. 454–61.

Ramsey, F.P. (1931). *The Foundations of Mathematics and Other Logical Essays*. Routledge and Kegan Paul.

Robinson, J.R.W. (1986). 'Life Cycle Costing in Buildings: A Practical Approach'. In *Australian Institute of Building Papers* (Volume 1). Paragon, Canberra, pp. 13–28.

Ruegg, R.T. and Petersen, S.R. (1987). *Comprehensive Guide for Least-Cost Energy Decisions*. US Department of Commerce, Washington.

Stone, P.A. (1973). 'Economic Criteria and Building Decisions'. In *Value in Building* (G.H. Hutton and A.D.G. Devonald, eds.), Applied Science Publishers, pp. 31–45.

Stone, P.A. (1974). 'The Technique of Cost-In-Use'. *Building* (London), vol. 227, no. 8, pp. 72–74.

Stone, P.A. (1975). 'Building Design Evaluation and Long Term Economics'. *Industrialisation Forum*, vol. 6, no. 3–4, pp. 63–70.

Stone, P.A. (1980). *Building Design Evaluation: Costs-in-use*. E. & F.N. Spon.

Stone, P.A. (1983). *Building Economy* (3rd edition). Pergamon Press.

Thomas, R. (1999). *Environmental Design: An Introduction for Architects and Engineers* (2nd edition). E. & F.N. Spon.

Treasury Department (1978). 'Discounting in Australian Public Sector Analysis'. Minute paper received by the Department of Construction, Canberra.

Tucker, S.N. (1987). 'Life Cycle Costing of Buildings'. In proceedings of Facilities Management '87 Conference, Melbourne, pp. 57–64.

Urien, R. (1975). 'Some Thoughts about the Economic Justification of Life Cycle Costing Formulae'. *Industrialisation Forum*, vol. 6, no. 3–4, pp. 53–62.

Willis, K.G. (1980). *The Economics of Town and Country Planning*. Granada.

21

Occupancy costs

Peter Smith

21.1 Introduction

The life-costs of a building represent the total cost of creating and maintaining it over a specified time horizon. Terminology commonly used to describe this total cost approach also include life cycle cost, recurrent cost, costs-in-use, operational cost, running cost, ultimate cost, terotechnology and occupancy cost. The variety of terminology reflects the different approaches and definitions associated with this area. However, many of these terms are too specialized or vague to adequately describe the total cost approach and debate regularly occurs over the correct definitions for these terms.

Occupancy costs fit into this category. For the purpose of this paper, occupancy costs are defined as operational costs that relate to a building's functional use, thus being only one, albeit important, element of a building's life-cost. These costs include staffing, manufacturing, management, supplies, etc. that relate to a building's functional use. Conflicting terminology is also used to describe this area with functional use-costs, operational costs, recurrent costs and running costs common descriptors.

In the construction industry, cost analyses of the operational stages of buildings almost exclusively focus on maintenance, repair, refurbishment, energy and other operational aspects of the actual building. The concept of occupancy costs requires buildings to be viewed from a very different perspective where the focus is placed on the end-user of the building.

21.2 Significance

Occupancy costs are often the most significant cost element in a building's life cycle. Nevertheless, these costs are normally the least subject to cost scrutiny during the design and life-cost analysis stages. Although an integral part of good design practice, functional-use analysis does not normally incorporate rigorous cost analysis due to it being the most difficult area to quantify. However, the following studies strongly suggest that, rather than being given token attention, it should in fact be the most important design and life-cost consideration.

A recent study of a CBD office building in Sydney over a 50 year life revealed the following life-costs: capital costs 11%, finance costs 7%, operating costs (energy, cleaning, maintenance, repair and replacement) 15% and occupancy costs 67%. Of the 67% relating to functional use, the largest component was clearly salaries (Hatzantonis, 1993).

BOMA (1991) carried out a national survey of large office buildings in the United States in 1990 and found that the cost of employees, an integral component of occupancy costs, pales other operational costs into insignificance. In terms of average annual expenditure expressed in US$ per square foot, operating costs were identified as: repair and maintenance $1.37, electricity $1.53, total energy $1.81, gross office rent $21.00 and office workers' salaries and associated costs $130.00. Staff salaries cost approximately 72 times the cost of energy (Romm and Browning, 1995).

These results are compatible with a study carried out by the Department of Housing and Construction (1980, p. 22): 'When compared with the total salaries bill for the occupants of the building, the initial capital cost and the running costs appear almost insignificant. For example, over the life of a typical office building for 50 years, studies have shown that 92% of the total cost is for the employees housed in the building, 6% is for running costs and a mere 2% for the capital cost.'

Service (1998) undertook a life-cost analysis for a typical hospital and came to similar findings for this specialist facility. Salaries and associated worker/occupational costs were the dominant factor. Worker cost categories were broken down into employee wages and salaries, employee benefits, professional medical services and contracted medical services. Together they accounted for 71% of life-costs. Associated work-related occupancy costs (medical supplies, drugs/pharmaceuticals, food and other medical expenses) accounted for a further 17%. The actual building capital cost accounted for only 6% of costs and the building operational costs also only 6%.

Worker/salary costs have not traditionally been included in life-cost analyses in the construction industry. The incorporation of this building life-cost element requires construction professionals to view buildings from a very different perspective – a perspective that ultimately focuses on actual building performance and the end-user.

21.3 Occupancy cost categories

Occupancy costs can be subdivided into acquisition and recurrent expenditure. The former comprises outlays for the initial purchase of goods and services related to the functional use of the building. Examples include office fit-out, stationery, computers and medical supplies. Denial-of-use costs, caused by construction delays in the acquisition of a building, can also be included under this banner.

Recurrent expenditure comprises the ongoing costs of maintaining the functional use of the building. Examples include staff salaries, manufacturing raw materials, washing and sterilizing expenses and the replacement of office supplies.

Hatzantonis (1993) identifies the following occupancy cost components as the most significant.

21.3.1 Salaries

Depending on building type and functional use, the largest cost factor influencing the overall cost of a project is often staff salaries, yet it is one that is normally ignored during the design stage. The design of a building can, in fact, influence the salary costs of staff using the building in four main areas: travel within the facility, supervision, work environment and worker productivity.

Travel and supervision are unproductive activities that are inherent in many operations. Therefore, minimizing these activities can lead to reductions in operation cycle times which can result in more productive work practices or even reduced staff numbers.

However, the design influence on worker productivity through improving the work environment constitutes the greatest scope for reducing a building's life-costs. In this context, productivity is measured in terms of improvements in production rate, quality of production and rates of absenteeism. The importance of the productivity area is such that it comprises the focal point of this paper.

21.3.2 Plant and equipment

Plant and equipment represent another major occupancy cost element. Examples include loose furniture, fixtures and equipment in an office building, photocopiers in a printing store and machinery in a factory. As these items are not normally integral with the building and are often purchased after occupation, building design has traditionally had little influence on their costs.

However, the increased use of loose-fit adaptable building designs has seen the interior design of buildings take on new dimensions. For instance, in office buildings, increasing focus is being placed on matching the ergonomic design of office furniture and equipment with the building design objectives. This can maximize the benefits of design concepts like daylighting and open space work areas that aim at improving worker productivity. Loose-fit workstation design is becoming an important part of the office building design process.

Additionally, building design can assist in maximizing the performance of plant and equipment. Hatzantonis (1993) cites examples such as the inclusion of varying voltage dampeners to the electrical installations of a building which will house delicate electrical equipment, analysis of the total wattage output or heat load of machinery in a building in order to ascertain air conditioning requirements or simply the correct choice of flooring to minimize wear on mobile equipment.

21.3.3 Operational supplies

The nature of operating supplies varies with the function of the building. Examples include paper in an office, bandages in a hospital, sheets in a hotel or food in a prison. Costs associated with this area are primarily affected by functional, organizational and management requirements rather than by building design. An exception, though, lies in the recycling and disposal of these consumables.

The incorporation of a recycling facility in the design of a building can reduce the ongoing cost of operational supplies. For example, glass works have facilities for purchasing, storing and recycling old glass products. The feasibility of such a facility will, of course, be dependent on the functional use of the building.

Waste disposal of consumables and the by-products of operations is becoming an issue of increasing concern as environmental standards and requirements continue to become more stringent. Waste disposal is another function-related occupancy cost. Most organizations produce some type of waste product requiring disposal. This leads to deterioration of the environment due to disposal through avenues such as exhaust, stormwater and sewerage systems as well as at dumping sites and tips. This has led to demands for stricter environmental protection legislation and a possible outcome might be a user-pays system in the future.

Hatzantonis (1993) contends that as we move towards user-pays systems, organizations might be obliged by law or find it more economical in the future to treat their own waste. Considerations of this type during the initial design might offer substantial savings later in the building's life.

21.3.4 Advertising

Although largely a management-based activity, advertising and marketing can be complemented by appropriate building design. External building signage or other design features can be used to increase public awareness of the organization. Similarly, the quality, reputation and/or image of a building can influence public perception and knowledge of its occupants. These measures can have a major influence on business performance through increased sales or market share. However, the net benefits can be difficult, if not impossible, to accurately predict or calculate.

21.3.5 Building alterations

Although building repairs and replacement are normally included as other forms of operating cost, building alterations that result from expansion of the function undertaken by a business might need to be anticipated during the design stage. The anticipation of and planning for future expansion or alteration in the initial design can greatly reduce operational costs when it comes time for the alteration work to be carried out.

21.4 Influence of building design

Although not all occupancy costs are affected by building design, certain design elements have a greater influence than others. Floor area is the most influential element. 'A reduction in floor area would not only contribute marginally to operating cost savings by cutting cleaning and repainting costs as well as energy costs for heating and cooling, but also lead to increased efficiency in a facility' (Green et al., 1986, p. 42).

Despite their potential significance, staffing costs have generally been viewed as being attributable to managerial and organizational factors but unrelated to building design. Green et al. (1986), however, reason that the design of buildings with highly specialized functions can greatly influence staffing costs. A hospital, for example, has very intricate and specialized design requirements where factors such as supervision and travel distances are important. If these can be optimized for each functional area then it is reasonable to assume that across an entire hospital the total time savings could result in a reduction in staff numbers. This may result in substantial savings in direct staff and administration costs. Even if no staff cuts are made, savings are still available through the minimization of unproductive time.

However, the greatest potential for reducing life-costs lies in enhancing worker productivity through improving the working environment within a building. The major difficulty though is determining the design areas that most influence productivity and the means by which any productivity changes are measured.

21.5 Worker productivity: case studies

The following collection of case studies, drawn from Romm and Browning (1994), are widely cited as prime examples of how good design can improve worker productivity. As will be seen, the major benefits arise out of 'green' environmentally-friendly design and particularly improvements in lighting and indoor air quality. The studies include both retrofits (refurbishments) and new buildings. Retrofits probably provide the best means of measuring productivity changes as the 'before and after' analyses are more readily comparable than is the case with new buildings.

Interestingly, the majority of these works and design changes were carried out solely to reduce energy, maintenance and repair costs. Subsequent improvements in worker productivity were an unexpected bonus that ultimately impacted most significantly on the bottom line of businesses (i.e. profit).

21.5.1 Reno Post Office, Nevada, USA

The Reno Post Office was renovated in 1986 with the sole objective of reducing energy costs. The retrofit design changes cost US$300 000 and were simply based on a lowered ceiling and improved lighting. The resultant energy and maintenance savings came to about US$50 000 per year which represented a payback period of six years. The major benefit though arose in the unintended area of worker productivity. A quieter and better lit work environment facilitated dramatic improvements in work output and quality. Productivity increases stabilized at 6% after one year and the mistake rate dropped to the lowest of any post office in the country.

21.5.2 Pennsylvania Power and Light, USA

This company experienced problems with their lighting system in a building that housed its drafting engineers. The main problem related to distracting glare and reflections off work surfaces. The lighting system was redesigned from general lighting to task lighting with high-efficiency lamps and ballasts. The cost of the renovation was minimal (US$8362) but the effect dramatic. Total energy and operating costs fell by 73% resulting in US$2035 annual savings. This represented a payback period of 4.1 years and a return on investment of 24%.

Based on the time it took drafters to complete drawings, productivity increased by 13.2% which was worth US$42 240 per year. This reduced the payback period from 4.1 years to 69 days and increased the return on investment to 540%. The quality and accuracy of work increased, the health of the workers improved, the sick leave rate dropped by 25% and worker morale skyrocketed. The value of reduced errors was calculated as being worth US$50 000 per year, which further increased the return on investment to over 1000%.

21.5.3 Wal-Mart, Kansas, USA

Wal-Mart, one of the largest retailers in the USA, developed a new prototype store called the 'Eco-Mart' in 1993 where focus was placed on ecologically sustainable design. Corporate objectives were primarily to foster an environmental image for the retailer as well as to reduce energy and maintenance costs. Although some of the design features were not entirely successful, daylighting produced the most startling, and unintended, results. To reduce costs, the skylights were installed in only one half of the building. Sales activity was significantly higher for departments in the daylit section of the building. Sales were also much higher than for the same departments in other stores. This, accompanied by an improved corporate image through marketing of the Eco-Mart concept, substantially increased sales activities and profit levels.

21.5.4 ING Bank, Amsterdam, Holland

The ING (formerly NMB) bank's headquarters in Amsterdam, completed in 1987, is commonly heralded as one of the most striking examples of how environmentally-friendly design can result in short payback periods, reduce operating costs, increase worker productivity and improve corporate image with the ultimate result of increasing market share and corporate profits. Construction costs were comparable to other office buildings in Holland but the additional energy measures resulted in estimated energy savings of approximately US$2.6 million per annum, which represented a payback period of only three months. The building used only one-tenth of the energy used in ING's previous building and only one-fifth of that used in a nearby more conventional office building constructed at the same time for roughly the same cost.

However, the greatest benefit lay in the resultant increase in worker productivity (absenteeism fell by 25%) and the bank's enhanced corporate 'environmentally-friendly' image. This has provided the catalyst for a dramatic increase in ING's business as it has moved from fourth to second in terms of the market share of financial institutions in Holland.

21.5.5 Lockheed Building 157, California, USA

The considerable benefits of energy-conscious daylighting design quickly became evident upon completion of Lockheed Missile and Space Company's commercial office building in 1983. Housing 2700 engineers and support people over a total floor area of 6000 m², a primary design objective was to reduce energy usage by at least 50%. Total annual energy costs were halved with daylighting alone reducing lighting costs by 75%. The energy-efficient improvements were estimated to have added US$2 million to the US$50 million construction cost. The energy savings were worth US$500 000 per year representing a payback period of only four years.

However, this was overshadowed by improvements in worker productivity. The average absenteeism rate dropped by 15% which, in itself, paid for the extra energy-efficient construction costs in less than one year. The architect, Lee Windheim, cited in Thayer (1995b, p. 28), contends that 'a mere 2% of the total cost of Building 157 over its useful life will be for the initial design and construction. Another 6% will be spent on maintenance costs, including energy, and employee salaries will account for the remaining 92%. This is why the contribution of the building to employee productivity is the most important design consideration. Small gains in productivity can make huge differences in corporate profits.'

The significance of productivity gains in increasingly competitive markets is further highlighted by the claim by Lockheed management that the improved productivity gave them the competitive edge that helped them win a US$1.5 billion contract. The profit made from this contract more than paid for the entire construction cost of their building!

21.5.6 West Bend Mutual Insurance, Wisconsin, USA

This insurance company's new 15 000 m² office building, completed in 1991, also has a number of energy-saving design features that have had a significant impact on operating costs and worker productivity. These features include an energy-efficient lighting system (task lighting and occupancy sensors), larger window areas, shell insulation, 'environmentally-responsive workstations' with individual temperature and airflow control and a more efficient HVAC (heating, ventilation and air-conditioning) system.

Whilst annual electricity costs were reduced by 40% the real benefits were found in improved worker performance. The 'environmentally-responsive workstations', the improved lighting and the improved thermal comfort of the occupants were all significant contributors to a productivity increase of approximately 16%. With an annual payroll of about US$13 million this represents a theoretical saving of over US$2 million per annum.

21.6 Healthy working environments

Absenteeism and reduced productivity due to ill-health are an important consideration when evaluating worker productivity particularly as the workplace itself can be a primary source of ill-health. Increasing evidence has been gathered over the past two decades to show that subtle problems associated with actual building design and function are having a significant effect on the well-being of inhabitants. This is commonly referred to as the sick building syndrome (SBS).

21.6.1 Sick building syndrome

SBS relates to 'illness associated with the indoor environment where symptoms are non-specific and their causes unknown' (Dingle, 1995, p. 19). Dingle contends that if the cause of the disease is known, such as in the case of legionnaires disease, it is called building related illness (BRI). Together they are having an alarming, and increasing, effect on worker productivity.

In the United States, it has been estimated that SBS costs approximately US$3 billion in lost annual productivity whilst in Australia it is estimated to cost industry several hundred million dollars each year (Dingle, 1995).

However, Dingle points out that SBS does not occur in all buildings but is normally associated with buildings used for non-industrial purposes and especially office blocks. Symptoms typically include irritation of the eyes, nose, throat and skin, mental fatigue, reduced memory, lethargy, headaches, dizziness, nausea and unspecific hypersensitivity reactions. Additionally, airborne diseases such as the common cold and flu are easily transferred via conventional air-conditioning systems.

21.6.2 Design factors associated with SBS

Dingle (1995) also found that research has consistently shown that the following design elements appear to increase the prevalence of SBS: wall-to-wall carpets, large areas of upholstery, large amounts of shelving and horizontal surfaces, a sense of crowding, office size, poor lighting, poorly designed workstations, air quality and poor environmental control by individuals. In terms of actual work practice, photocopying, handling carbonless copy paper and working with a video display unit all appear to contribute to SBS. Broadbent (1993) also cites psychological and social factors as contributors. These factors may manifest themselves in the stress associated with new technology and work practices.

The most significant determinant appears to be the heating, ventilation and air-conditioning (HVAC) design. '50% of problems are due to poorly designed, operated or maintained heating, ventilation and air conditioning systems' (Dingle, 1995, p. 21). Studies have consistently showed that these problems can be significantly reduced if a building incorporates natural ventilation in all or part of its design (Rowe and Wilke, 1994). In a survey of 10 Australian office buildings, ranging in age from 5 years to over 100 years, Rowe and Wilke found that the 100 year old building, one of only two naturally ventilated buildings surveyed, easily produced the best results in terms of air quality and lower susceptibility to SBS. The other naturally ventilated building ranked second.

Good maintenance and frequent effective cleaning are also important. Poor cleaning equipment and practices can lead to elevated levels of dust in the air and a corresponding increase in symptoms. The maintenance and cleaning of HVAC systems is clearly another critical factor.

21.6.3 Ergonomic factors

Broadbent (1993) has also found that the interaction between the individual worker, the design of their workplace and the demands of their job has an effect on health. Key design features in this area include lighting, workstation design, interaction with other workers and ergonomic design of office furniture and equipment.

21.6.4 Ramifications

The ramifications of a poor work environment do not only relate to reduced productivity levels. Building proprietors and managers are increasingly concerned about the increase in workers' compensation costs and the potential legal consequences of failing to adequately address SBS and other building-related factors that impinge on the health of workers. The future potential costs in this area alone may lead to financial ruin.

21.7 Key design influences on productivity

These case studies strongly suggest that, in terms of improving worker productivity, visual acuity, thermal comfort and air quality are key determinants. Daylighting and natural ventilation are design concepts that can have a tremendous influence in this area. Clearly, workers who are comfortable and enjoy their working environment are likely to work more efficiently and productively.

Lippiatt and Weber (1992) found that building design affects productivity through ambient conditions such as noise, air quality, lighting and temperature and through workplace conditions such as enclosure, size and layout. However, they express concern at the lack of reliable and adequate data and research in this area. Research is clearly needed to assist designers.

To date, the Buffalo Organization for Social and Technological Innovation (BOSTI), cited in Brill (1984), has probably undertaken some of the most significant research and compiled one of the largest databases in this area. BOSTI undertook a five-year research programme to quantify design impacts on worker productivity which involved over 6000 employees in 70 organizations. BOSTI identified six office features that most significantly affect productivity: temperature fluctuation, air quality, glare, noise, enclosure and furniture. The effects of improvements and declines in these features on productivity were assessed but productivity impacts for specific levels of change were not published.

21.8 Applicability of occupancy costs

The case studies demonstrate that occupancy cost analysis is ideally suited for buildings procured for business or work-related purposes. In these circumstances 'buildings' would be better viewed more as 'dynamic business facilities' where the focus is on the functional needs and performance of the actual users. The quality of user/worker performance has traditionally been attributed to individual, managerial and organizational factors but the case studies show that building design can have a profound effect on user/worker performance.

Green et al. (1986) provide another twist to this argument. They contend that the design of buildings with highly specialized functions can greatly influence staffing costs. A hospital, for example, has very intricate and specialized design requirements where factors such as supervision and travel distances are important. If these can be optimized for each functional area then it is reasonable to assume that across an entire hospital the total time savings could result in a reduction in staff numbers. This may result in substantial savings in direct staff and administration costs. Even if no staff cuts are made, savings are still available through the minimization of unproductive time. Nevertheless, the greatest potential for reducing life-costs lies in enhancing worker productivity through improving the working environment within a building.

In these circumstances, the building is actually a vital business tool, not simply a dormant product. This is now being recognized by businesses and governments worldwide. Managers and organizations are increasingly incorporating the quality of the work environment created by a building's design as a key element in improving work/business performance and increasing profits. This requires designers to place more emphasis on the effect of their designs on the actual performance of the end-users. Building designers have a key role to play in the future business performance of the end-users and this is where the greatest value to the proprietor may well lie.

However, many buildings are not procured for the purposes of owner-occupation, where occupancy cost analysis is most beneficial. The greatest value for developers intent on a quick sale will usually lie in maximizing profit through maximizing the difference between total capital cost and selling price.

Developers in this situation are far less concerned with operational costs and the end-user.

The proprietor who retains ownership of the building upon completion with the intention of leasing will clearly be more concerned with operational costs particularly with respect to energy, maintenance and repair costs. However, these proprietors are not primarily concerned with the end-user, the tenant or lessee's operations. At the design stage, they may not even be able to readily identify the type of activities that will be undertaken by future users.

Nevertheless, building occupiers are become more astute and aware of the effect of building design on business/worker performance. More attention to occupancy cost analysis may enable these proprietors to use the resultant benefits as a marketing tool to increase the selling price or lease price, thereby increasing the value of the building to not only the immediate proprietor but also to future owners and users.

Due to the typical significance of salaries, occupancy costs may not be an important consideration for buildings whose occupants are not in income earning positions (e.g. residences, community centres). Strictly financial occupancy analyses are suited to buildings procured for government, office, retail, manufacturing, services and the like.

However, the process of occupancy cost analysis during the design stage can be used to influence other intangible benefits that are difficult to express in monetary terms but nevertheless lead to better value for money. Building designs that produce optimum environments for end-user performance can, for example, enhance student learning in educational facilities or improve the recuperative rates of patients in a hospital. Additionally, productivity gains are not always measured in terms of work output and quality. Better building design can lead to improved business performance through increased sales and improved corporate image or reputation.

21.9 Constraints

Despite all of this, the ability of building design to influence occupancy costs and particularly worker productivity has generally been met with scepticism in the construction community. A major problem is that it is not so difficult to demonstrate retrospectively that 'good' design which focuses on reducing functional-use costs makes good economic sense but it is often very difficult to determine how to ensure good design before it has started (Bon, 1996). Although considered a 'hot' topic now among facilities/asset managers (Hannus, 1996), it remains a relatively new area that currently relies on largely anecdotal information rather than a sufficient volume of accurate historical data and research. Other commonly cited problems include the following.

21.9.1 Economic/market realities

Difficulties associated with predicting into the future will always present problems. Moreover, the total life-costs and occupancy costs of a building are rarely borne by the one organization. The majority of building developers in Australia are not owner-occupiers and are typically not concerned with the long-term operational costs of the building. The same applies for people who buy or lease the building from the developer. Compounding all of this is that traditional discounting techniques are misleading in that they render long-term operational costs inconsequential in relation to earlier ownership costs.

Nevertheless, attitudes are changing as governments, building proprietors and tenants become increasingly aware of the enormous costs associated with occupying, operating and maintaining a building.

21.9.2 The Hawthorn Effect

Research carried out in the 1920s at Western Electric's Hawthorn plant in Chicago found that 'contrived experiments to monitor the effect of workplace change on productivity can be complicated by the special conditions of the experiment' (Browning, 1996, p. 2) and that any gains are only temporary. This became known as the Hawthorn Effect, which implies that changes in the physical work environment only have an effect on worker productivity because:

1. Workers appreciate the concern and interest of management.
2. Workers know that their productivity/absentee rate is being monitored.
3. Workers may be given bonus incentives for improved productivity.
4. There is often a special interaction between the worker and the researcher and the conditions of the experiment may favour the desired outcome.

The Hawthorn tests were, however, flawed in that the research was only based on five people and their supervisors monitored their own production rate and were rewarded for any gains in productivity. Nevertheless, 'the Hawthorn research led to widespread belief that changes in working conditions affect productivity only because they signal management concern. The effects of building design on productivity were basically ignored for 60 years' (Browning, 1996, p. 2).

Dickson and Roethlisberger (1986) found that this traditional view of the Hawthorn Effect was very different from the actual researchers' findings. They concluded that productivity can be enhanced by a more co-operative relationship between management and labour, a greater identification by workers with the goals of management and management effort to treat workers with respect and to be responsive to their needs and abilities.

The greatest influence on worker productivity comes from good management and organizational practice. However, ensuring the optimum working environment should be an integral part of this practice is where building design has a significant role to play. Studies by Brill (1984) and Romm and Browning (1994) demonstrate that the close links between design and productivity cannot be ignored by organizations intent on improving worker productivity and overall performance.

The Reno Post Office case study provides perhaps another reason for the persistence of the Hawthorn mythology. The results were originally never published because the productivity gains were not due to any changes in management practice for which the managers could take credit. However, the managers eventually convinced themselves that it was, in fact, due to their management because it was their decision to do the retrofit. The fact that the productivity gains were not planned nor anticipated was pushed into the background.

This raises the issue that managers have a vested interest in productivity gains. The illusory or fostered perception that productivity gains are due primarily to good management practice rather than good design have perhaps stymied the recognition of the important role that good design has on productivity.

Worker productivity is intertwined between work environment and management. Whilst the Hawthorn Effect is a legitimate theory, the reality is that good design and the Hawthorn Effect compliment each other. Improving the work environment does indicate management interest and concern. Design that focuses on a building's functional-use requirements and the needs and wants of the end-users, the occupants, all makes good management sense. Involvement of employees in the design process can lead to more accurately targeted design, increased staff morale and esteem and genuine desire for the design initiatives to be effective. As Dickson and Roethlisberger (1986) point out, staff should also be made aware of management goals and objectives.

Good design can provide the catalyst for all of this and, combined with good management practice, can lead to sustainable productivity gains.

21.9.3 Productivity design knowledge

Hatzantonis (1993) suggests three basic reasons why staff salaries are not normally considered at the design stage. Firstly, there is a misconception that design cannot influence expenditure on staff salaries. Secondly, other life-cost components can appear insignificant compared to staff salaries. Lastly, a high level of knowledge of the functional uses of buildings needs to be known by the design team. Circulation, work patterns, processes and even habits will all have an impact on how the function within the building is undertaken by staff. Additionally, knowledge of design features that will produce an optimum work environment to facilitate productivity maximization is necessary.

This last point is the largest inhibiting factor. The modern-day designer needs to consider a plethora of variables when contemplating 'green' design. The economic reality is that designers generally do not have the time, finances or resources to carry out sufficiently in-depth research to establish the most effective 'green', energy-efficient and productivity-enhancing design features. To start with, these will vary according to building type, functional use and client requirements.

As mentioned previously, further research is required to assist designers by providing sufficiently detailed and reliable data and information. This needs to be combined with general education and awareness programmes to spread the knowledge gained and provide the necessary evidence required to sell this design concept. Implementation will be more quickly facilitated through statistically rigorous data and research rather than ad-hoc and largely anecdotal information.

Due to the wide variety of variables impinging on this area, research ideally needs to be based on a team approach incorporating input from building owners, architects, engineers, government bodies, services authorities, environmental experts, energy experts, construction cost experts, facilities managers, employees and even sociologists.

The emergence of facilities/asset management as a discipline in its own right has seen a marked increase in the level of post-occupancy evaluation studies. The incorporation of analyses of the effect of design on worker productivity, currently given only scant, if any, attention could provide the most practical means of building up the necessary data and information.

21.9.4 Obsolescence

The future obsolescence of building designs has always been an important consideration in life-cost appraisal. Advances in information technology (IT) have heightened this importance as many work practices undergo radical change leading to redefinitions of the 'workplace'. This poses great difficulties for occupancy cost analysis as it requires an emphasis on establishing current and future end-user (worker) requirements.

Current design needs to consider the changing needs of the work environment as well as changing lifestyles. With IT developing so rapidly, it is difficult to forecast requirements in two to three years let alone over the anticipated life of a building. Loose-fit adaptable design is gaining increasing importance as the most logical means of addressing this problem.

Bautista, cited in Ketchell (1995), provides a good example of these changes. She comments that a trend sweeping the United States is for office space to be cut by up to two-thirds by strategic facility consultants which can save millions of dollars from a company's bottom line by drastically reducing upfront and operational costs. Focus is shifting from large individual offices to high-performance collaborative team work space. This has necessitated a different way of rewarding seniority and status within the workplace which stretches beyond the realm of who has the largest desk and office. It is predicted that this trend will have a huge impact on office space utilization worldwide.

Telecommuting is another example. IT advances have reduced the need for workers in certain sectors to be within a central workplace. Building design emphasis in these sectors may, in time, shift from a central office or workplace to the workers' individual homes. This can drastically reduce overhead costs and studies have shown that workers can work more productively from home. This could actually lead to a re-think about the average house design to ensure that telecommuters have an appropriate working environment. When one considers that the motor car is one of the greatest polluters in our environment, the greatest advantage with telecommuting from an environmental perspective lies in the reduction of worker travel requirements.

As Australia already has one of the highest rates of people working from home in the world, telecommuting is an important consideration that will require a fundamental shift in thinking both from a business and design perspective.

21.9.5 Measuring productivity changes

The case studies have shown that even minor productivity changes can have major cost consequences. Lippiatt and Weber (1992) found that, on an annualized per square metre basis, employee salaries in office buildings are approximately 13 times building costs. Theoretically, a 13% increase in construction costs for productivity-enhancing design can be justified if it increases productivity by 1%.

However, there must be a limit on the effect that design can have on productivity. Ward (1987, p. 115) surveyed research on the impact of building design on productivity and found that productivity gains of up to 10% are more than feasible. He highlights the potential global significance of this by stating that 'given the US annual expenditure of $1 trillion for white collar salaries a 5% productivity increase translates into US$50 billion per year'.

This, however, can legitimately be treated with scepticism as the accurate measurement of productivity changes remains undeveloped. Lippiatt and Weber (1992, p. 1) contend that 'current literature reveals heightened awareness of the importance of productivity impacts on building design yet little progress has been made toward systematically including productivity impacts in building economic analysis. Without explicit treatment of productivity, it is difficult to justify higher-priced designs on productivity improvements'.

Lippiatt and Weber (1992, p. 18) also note that despite the fact that salaries often far outweigh any other life-cost element, life cycle costing does not account for the productivity benefits of new technologies and building design. It traditionally judges project alternatives on building cost-effectiveness alone. 'Higher priced designs that enhance productivity may make more economic sense. Yet without an economic method for systematically including productivity benefits in building Life Cycle Cost analysis, these designs cannot be justified.'

Accordingly, they have developed two methods to facilitate such analysis but admit that more comprehensive supporting data relating building design to productivity is required to effectively implement them. The first is described as the net benefits (NB) method which expands the life-cost approach to incorporate not only the costs but the benefits of design alternatives thus measuring economic efficiency rather than cost-effectiveness. In the private sector, economic efficiency is synonymous with maximum profits whilst in the public sector this often relates to maximum net benefits. The method requires monetary measures of productivity benefits and, to this end, utilizes a Productivity Impact Function (PIF) system that measures 'cause and effect' relationships between building design and productivity.

The second method, multi-attribute decision analysis (MADA), also incorporates worker productivity in life-cost analysis. This technique enables more than one performance or design attribute to be considered in a decision even if the attributes are not measured in comparable units. It simultaneously accounts for both traditional life-cost measures in dollars and non-monetary measures of productivity impact. Productivity impact may be expressed in physical dimensions, such as square metres or decibels of sound energy, or even be based solely on informed judgements.

However, an added difficulty is that the economic evaluation of design proposals should involve an estimation of the value of staff salaries. Workload and salaries are the two main components that need to be considered. In terms of workload, the number and nature of the workers will be required. The estimated number of hours per worker will be the basis for calculating a total cost for staff salaries. If workloads are built-up from the hourly requirements for various activities it is important to make allowances for sickness, leave, etc.

Salaries are another crucial area that must be approached with caution. The type of worker, the nature of the work and the profitability of the company that will occupy the building will all have an effect on the level of salary. As employee salaries can amass to a significant proportion of a project's total cost, care must be taken in both the calculation and the presentation of any such forecasts. Additionally, a comprehensive sensitivity analysis should be undertaken on the various factors influencing the cost of staff salaries.

The accurate measurement and prediction of the effect of design features on worker productivity is critical to the acceptance of this concept. The greater the certainty of outcome and benefits, the more willing building proprietors will be to embrace this design concept.

21.10 Key selling points for green design

It has been established that productivity-enhancing design features predominantly stem from 'green' design. Consequently, the ability to accurately demonstrate the effect of productivity benefits on payback periods and profit levels will potentially be the greatest selling point for environmentally-friendly and ecologically sustainable design.

21.10.1 The 'hip-pocket principle'

Societal support for environmental protection and conservation has, particularly in developed countries, increased rapidly over the past decade and will continue to do so. This has coincided with increased publicity and education about the problems besetting our planet resulting in widespread awareness of the importance of environmental protection.

However, this support has its limitations. The reality is that, hard-core environmentalists aside, the average person or business will only be as 'green' as their finances comfortably allow. Relying on the ideals and goodwill of people to embrace environmental measures to protect future generations is unrealistic. The ultimate determinant of how environmentally conscious and active people will be is how far they will need to dip into their pockets for extra cash (the hip-pocket principle).

Solutions to our environmental problems need to be economically based; they need to demonstrate that they will save people/business money or make them money. Unfortunately, in the real world, it is money and not values or ideals that drive the future direction of our planet. Strategies need to be balanced with the short-term impact on people's hip-pockets. Solutions that ignore this principle are doomed to failure.

Governments normally provide the catalyst for change but they are also constrained from doing so by the hip-pocket principle. While it is important for many governments to at least appear 'green' they cannot implement environmental policies that will impact too greatly on hip-pockets. The prime movers for ecologically sustainable development are basically democratic governments from developed countries and, to stay in power, have to be wary of measures that hurt people or business financially.

The principle is even more relevant with developing countries. Their primary goal is generally economic development as they attempt to reach the levels attained by developed countries and this goal cannot be denied. The reality is that they often cannot afford or are generally not interested in controlling development or implementing environmentally-friendly solutions. The major challenge on a global basis is to come up with solutions that reduce costs or make money.

Consequently, this is why the concept of productivity-enhancing design has the potential to be the greatest selling point for environmentally-friendly design. Appropriately designed buildings and energy systems can substantially reduce operational costs, increase productivity and, most importantly, increase profits. The fact that this can best be achieved through environmentally-friendly design may provide an indirect, but most effective, means of implementing sustainable design solutions.

It must be noted that many buildings are not income-producing. However, intangible non-pecuniary benefits associated with improved working and living environments, such as aiding the learning process in schools and recuperative rates in hospitals, have been well documented (Thayer, 1995a; Ravetto, 1994). These benefits, whilst not necessarily reflected in strict monetary terms, may lead to improvements that people may place more value on than would be the case in a purely financial analysis. Additionally, the widespread implementation of green design may lead to more awareness and acceptance of the benefits that it may confer.

Perhaps the initial catalyst for change lies with retrofits of both existing public and private buildings. The case studies have indicated that benefits can be more readily demonstrated with retrofits. In both public and private sectors, there have been increasing concerns about the cost of operating and maintaining building assets and the effect that this has on dwindling operational budgets. The growth of facilities/asset management is largely in response to these problems. The fact that many environmentally-friendly retrofits can be carried out with extremely short payback periods and impressive long-term benefits must be seen as a tremendous opportunity to encourage green design.

21.10.2 Financial incentives

Rather than through direct regulation, governments could further encourage environmentally-friendly design via appropriately targeted building design tax concessions or subsidies. A starting point may be increased depreciation rates for adaptable loose-fit design and key design features such as building envelope and energy systems that reduce energy usage and improve air quality. For example, 'if energy use is 10% below target levels then the allowable tax deduction may be increased to, say, 110%' (Best, 1996, p. 4). This will further reduce payback periods and operational costs for building proprietors.

21.11 Conclusion

The concept of occupancy costs, whilst not new, remains in an embryonic state in terms of quantification and economic evaluation. Although potentially the most substantial life-cost component they are rarely considered in appropriate detail during the design process largely because of the lack of reliable data and economic measures to accurately predict the net benefits of various design alternatives. Perhaps the construction economist has an important role to play here.

Occupancy costs are now a leading topic in the facilities/asset management field due to the increasing awareness of the enormous costs associated with occupying, operating and maintaining buildings. Their importance cannot be denied but clearly more research is required before occupancy cost analysis can receive widespread acceptance.

Critics of occupancy cost planning would argue that all the matters raised herein are merely examples of good design. It is the designer's responsibility to examine the functional requirements of the proposed facility, to effectively plan the work environments, minimize circulation and cater for the future needs of the client. But the question remains, what is the true cost of these design features and do they represent optimum value for money when examined using a life-cost approach?

References and bibliography

Best, R. (1996). 'Controlling Energy Demand in Australian Buildings: The Carrot or the Stick'. Unpublished Paper, University of Technology, Sydney.

Best, R. and de Valence, G. (1999). *Building In Value: Pre-Design Issues*. Arnold.

BOMA (1991). *Experience Exchange Report*. Building Owners and Managers Association, USA.

Bon, R. (1996). 'Occupancy Costs'. E-mail Correspondence, University of Reading, UK.

Brill, M. (1984). *Using Office Design to Increase Productivity, Workplace Design and Productivity*. Buffalo, USA.

Broadbent, C. (1993). 'Sick Buildings: Preventative Strategies'. *Specifier*, February, pp. 54–56.

Browning, W.D. (1996). 'New Case Studies'. In proceedings of Green Building Now – Tools for Sustainable Design and Construction, 7–8 June, California, USA.

Department of Housing and Construction (1980). *TI140 AE – Life Cycle Costing: Technical Information*, Sydney.

Dickson, W.J. and Roethlisberger, F.J. (1986). *Counselling on Organization: A Sequel to the Hawthorn Research*. Harvard University Press.

Dingle, P. (1995). 'Sick Building Syndrome Defined'. *BEST*, November, pp. 19–21.

Ferguson, A. (1996). 'Time for the Office Building to Lift its Productivity too'. Business *Review Weekly*, vol. 18, no. 41, pp. 136–37.

Green, J., Adams, A., Nelson, S. and Aisbett, K. (1986). *Evaluating Hospital Ward Designs In Use*. School of Health Administration, University of New South Wales, Sydney.

Hannus, M. (1996). 'Occupancy Costs'. E-mail Correspondence, VTT Building Technology, Finland.

Hatzantonis, S.J. (1993). 'Recurrent Costs: A Role for the Quantity Surveyor'. Undergraduate Thesis, University of Technology, Sydney.

Ketchell, M. (1995). 'There'll be Little Room, No View at the Top'. *The Weekend Australian*, 2 December.

Kleeman, W.B. (1991). *Interior Design of the Electronic Office*. Van Nostrand Reinhold.

Lippiatt, B.C. and Weber, S.F. (1992). *Productivity Impacts in Building Life-Cycle Cost Analysis*. US Department of Commerce, Gaithersburg, USA.

Patterson, M. (1996). 'Investing Success'. *Buildings Online: The Magazine*. <http://www.buildingsonline.com>

Peck, M. (1993). 'Role of Design Professionals'. *Building Owner & Manager*, February.

Ravetto, A. (1994). 'Daylit Schools'. *Solar Today*, March/April, pp. 22–24.

Romm, J.J. (1995). 'Keep your Facility Fit with Lean and Clean Engineering'. *IIE Solutions*, June, pp. 17–20.

Romm, J.J. and Browning, W.D. (1994). *Greening the Building and the Bottom Line*. Rocky Mountain Institute, USA.

Romm, J.J. and Browning, W.D. (1995). 'Energy Efficient Design can Lead to Productivity Gains that Far Exceed Energy Savings'. *The Construction Specifier*, vol. 48, no. 6, pp. 44–51.

Romm, J.J. and Browning, W.D. (1996). 'Greening the Bottom Line: Increasing Productivity through Energy-Efficient Design'. In proceedings of Green Building Now – Tools for Sustainable Design and Construction, 7–8 June, California, USA.

Rowe, D. and Wilke, S. (1994). *Sick Building Syndrome and Indoor Air Quality: Perception and Reality Compared*. Department of Architectural and Design Science, University of Sydney.

Service, B. (1998). 'NSW Commission Leads to Performance-Based Contracts'. *Chartered Building Professional*, pp. 10–11.

Thayer, B.M. (1995a). 'A Daylit School in North Carolina', *Solar Today*, November/December, pp. 36–39.

Thayer, B.M. (1995b). 'Daylighting and Productivity at Lockheed'. *Solar Today*, May/June, pp. 26–29.

Ward, R. (1987). 'Office Buildings Systems Performance and Functional-Use Costs'. In proceedings of the Fourth International Symposium on Building Economics, vol. A, Working Commission W.55 on Building Economics, International Council for Building Research Studies, Copenhagen.

PART 8

Asset management

Sustainable practices are relevant to the pre-design (or strategic planning) and design (or concept development) stages of facility creation, including the construction process itself. But the commissioning of the facility marks the beginning of a long period of occupation that demands continual supervision and guidance. Known generally as asset management, it is a post-design activity that comprises operational issues arising from the functional-use of the facility over its effective life.

Operational issues involve making decisions on energy management, cleaning, maintenance, replacement of worn or obsolete components and other administrative matters. It encompasses gathering feedback about performance, making comparisons with designed criteria, and instigating change as new circumstances dictate. The expansion of asset management to also include strategic business planning aimed at improving productivity and competitiveness has led to the emergence of a new field called facility management.

The post-design stage provides opportunities for change and renewal, and so incorporates aspects of strategic planning and concept development applied to existing infrastructure. For example, when a business needs to expand its operations or branch into a new market, it will create implications for its existing facilities such as acquisition of new premises, refurbishment and extension, or demolition and new construction. Clearly post-design activities must take account of sustainable practices in much the same way as new development.

Asset management is as much about the management of change as anything else. The availability of new technology, for example, can make existing design solutions obsolete and generate opportunities to improve current performance, conserve energy, minimize waste and the like. Despite the long life of many building components and systems, new solutions that have significantly better performance can lead to the premature replacement of existing solutions. Re-evaluation is an ongoing process but otherwise no different from the procedures used for new development. The pursuit of sustainability should be a life-long activity for all professionals associated with the built environment.

Notwithstanding the introduction of new technology, existing facilities offer enormous potential for improvement. Performance measurement, regular monitoring and comparison against original design criteria can identify aspects of operation that are not functioning as expected. This may be a result of incorrect installation, age, component failure, inappropriate usage, lack of proper servicing or simply a poor decision in the first place. Corrective action can restore performance to match the original expectation and save further resource wastage.

Energy usage is a good example of the importance of continual monitoring and management. High energy consumption may require a rethink of current practices, reconfiguration of controls, installation of new control systems or even an education campaign about the merits of energy saving practices. Most facilities offer ways in which energy savings can be made. An energy audit is one method of identifying areas of high consumption, benchmarking performance against other facilities of similar function and recommending remedial action. While this will entail new expense, over a small payback period the cost savings can be significant. The contribution at a community level to reduced energy demand can also lead to social benefits and less environmental damage.

How do we know that our design decisions are effective and are contributing to sustainable development goals? The answer is not simply proper initial evaluation procedures. It is critical that monitoring of performance and continual feedback assist to review design decisions and to rectify them if necessary. Feedback is the final step in the quest for sustainability. Not only can feedback lead to improvements in the performance of existing facilities, but it can improve new designs for other projects and ultimately create standards against which future designs can be compared.

Post occupancy evaluation is a recognized method of learning about the success of recently completed facilities. It involves a combination of objective data collection and user opinion to identify aspects of operation that are sub-optimal. Unfortunately it is not a routine activity, but where commissioned it tends to occur after one year of occupation. The rationale for this is that four seasons would have taken place and therefore the impact of summer and winter climates can be observed. However, there is a convincing argument for more frequent occupancy evaluation. This may be a requirement in the future if facilities must satisfy operational benchmarks to attain initial planning consent. Clearly without some form of review and enforcement, planning criteria would be subject to abuse.

A process of self-assessment and annual reporting, perhaps linked to the taxation system, may be necessary to control future resource consumption (particularly energy) and ensure that facilities continue to operate as planned. Such a process may comprise periodic audits and penalties for non-compliance. If applied to all built facilities, both new and existing, it would raise sustainability to another level. Yet whatever the future holds, it is clear that treating sustainability as a one-off activity during project acquisition fails to guarantee success.

The chapters in this part concern post-design activities and the need for continual review and change. Chapter 22 advances the argument that post occupancy evaluation is a vital feedback mechanism and needs to be used to its full potential. Chapter 23 discusses the importance of environmental auditing, and in particular energy auditing, as a means for improving operational performance. Chapter 24 concludes by examining the emerging discipline of facility management and the integration of infrastructure management to core business activities.

The end is but a new beginning. The effective management of existing facilities is a combination of new opportunities and new decisions. In this context, strategic planning is not a pre-requisite to facility design and management, but integral with it. Sustainable practices relate to all aspects of the development and usage of infrastructure. They are multidisciplinary, and vital to the prosperity of society both now and long into the future.

22

Post occupancy evaluation

22.1 Introduction

Often building users complain that their workplace is not designed so as to suit their requirements and meet their needs. Controversially, building designers assert that their designs have carefully considered the users' needs and accommodated all of the specified requirements. Unfortunately the problem is often one of misunderstanding on both sides. It is true that designers have comprehensive knowledge of building design but usually receive little or no feedback after the building is complete. Today building technologies are progressing rapidly and buildings are becoming more complex and sophisticated. New constraints such as efficient resource usage and low environmental impact require innovative solutions. Without the opportunity to learn from past projects, designers may repeat solutions on new projects that have proven unsatisfactory in practice.

The evolution of post occupancy evaluation (POE) helps to bridge the gap between designers and end-users. POE is recognized and valued as a process that can improve and help to explain the performance of the built environment. It is a new tool for building owners and facility managers to maintain the quality and performance of buildings for which they are responsible. POE came into existence in the 1960s but is still a young field that requires further research and development. Feedback from POE is constantly quoted by building owners and facility managers as a common pre-requisite for the design phase in the building delivery cycle.

22.2 Building performance evaluation

In the past, building performance evaluation has been practised in the construction industry but in a simple and informal manner. This type of evaluation was perhaps sufficient at that time because building projects were relatively less complex and the level of technology in design was low. Lessons learnt and knowledge gained was able to be passed on from generation to generation. But things have changed dramatically in recent years. Building performance evaluation has to be improved to cope with the ever increasing proliferation of specialization in the construction industry in terms of matters such as building types, services, technology, code and regulatory requirements, energy conservation, fire safety and environmental health and safety constraints.

Construction is a highly sophisticated industry that has evolved from simple residential homes to complex projects incorporating high technology services and communications. The research and development of new technologies and its introduction into buildings is progressing quickly. Monitoring of these new technologies is necessary in order to compare actual performance with intended criteria, and through this process improvements can be found and incorporated in future building design. The consideration of environmentally-friendly techniques and selection of building materials with low embodied energy is particularly important. Energy efficiency technologies are still in a state of development and so data collection about their performance is crucial.

Building performance evaluation acts as a tool to record and compare actual performance with explicitly stated criteria. The traditional methods of evaluating building performance are inadequate to cope with highly complex construction techniques. In order that past experience can be passed on and incorporated in future planning, performance records need to be articulated and documented in a more systematic and objective manner. POE is a process that can capture the performance of the built environment and the accumulated information can form the basis for developing design guidelines and criteria for planning future buildings of similar types.

POE is an ideal tool to tackle significant problems experienced in building performance with particular emphasis on the perspective of the building occupant, as was first noted in institutional care buildings such as mental hospitals, nursing homes and correctional facilities (Preiser, 1995). Since then POE has been used widely by government bodies as well as private organizations ranging from projects which are small to very large in size. POE has developed from rather simple case studies of individual facilities to sophisticated studies of building types with valid and reproducible results.

POE forms a systematic way of comparing actual building performance with stated performance criteria since it focuses on user satisfaction. POE is a logical method of evaluation and can be readily applied to different situations in an attempt to evaluate, for example, a building, a floor or even a single workstation as part of ongoing day-to-day facility management. The intention of POE is to measure the fitness for purpose for a period after building construction activity has been completed and after the building has been occupied for at least a full year. This ensures that users will have experienced the building through different seasons under a variety of climatic conditions.

22.3 Performance evaluation processes

22.3.1 Role and purpose of POE

POE can be used for different purposes. It is up to building owners or facility managers to identify the purpose of an evaluation before selecting a suitable approach. POE provides an opportunity for management to consult users (including employees) about the workplace so as to improve their relationship for any future co-operative activity. The process of consultation emphasizes the importance of users in the overall management of an organization and, more importantly, users may feel the presence of a channel through which they can communicate problems or dissatisfaction to higher levels. This can improve mutual understanding and productivity. Nevertheless consultation raises expectation and it is important to ensure that visible results are made or else users will lose confidence, and participation in future evaluations will be difficult to regain.

POE can be used to assist with upgrading an existing facility. It is often very useful to conduct a POE before planning for any refurbishment or renovation because users' expectations, requirements and needs are addressed and the design team is properly informed. Similarly, POE responds well to staff complaints, dissatisfaction and understanding expectations. However, it is crucial to find out the underlying source of dissatisfaction before spending money imposing solutions that in the end turn out to have addressed the wrong problem. If dissatisfaction concerns the work environment, POE can be a useful tool in gathering information from users through the operation of a planning workshop. POE then becomes a suitable technique whereby designer, user and management levels are brought together to exchange information. Furthermore, POE can gather information that can be accumulated and developed into design guidelines, handbooks and other reference material for future design purposes.

22.3.2 POE models

In the POE process three main groups of people are involved. They represent different levels and styles of involvement, but each group is important and without their co-operation the POE process will never be successful. The groups are initiators, evaluators and participants.

The management level is always the initiator of a POE process for a building and has the role to engage an organization or person to carry out the evaluation. Management normally does not take part in the evaluation activities, but instead plays an important role in examining the results and implementing recommended measures to improve deficiencies.

An organization or person authorized to carry out an evaluation is called the evaluator (also known as the facilitator). This often comprises a group of experts who conduct on-site surveys and measure and record building performance. In some cases this role may be played by the facility manager, and so may have a dual role as an initiator of the process. Where a team approach is appropriate, it is drawn usually from a range of disciplines involved in the original design process.

Finally, the most important group in the POE process is the participants. They are the end-users who supply information about the building. They may represent different stakeholders and hence may comprise occupants, visitors, employees, customers or tenants. During the POE process they are arranged usually into small groups sharing the same interest. The participants may be partially or fully involved in the POE process depending on the time scale and purposes for carrying out the POE.

There may be three different levels of effort in the POE process (Preiser, 1989). The selection process depends upon the availability of finances, time, manpower and the required outcome. Each level consists of the same general approach of planning the process, conducting the study and interpreting the results:

1. *Indicative POE*. This is the most common form of POE and can be completed with a few hours of on-site data gathering. It comprises a quick walk-through assessment carried out by an experienced evaluator. Due to the very short time span, evaluators may have to make professional judgements themselves. The activities involved include data collection, archival document evaluations, structured interviews with key personnel, group meetings with end-users and general inspection of building performance. The findings are generally limited to providing an indication of major successes and failures of the building's overall performance and making direct assessments of the elements and functions in the form of a written report.
2. *Investigative POE*. Investigative POE is a more in-depth study of building performance. It may be required as a result of problems identified during an indicative POE. This can take a week to several months to complete depending on the depth of investigation and the amount of personnel involved. The usual techniques are scheduled interviews and survey questionnaires. This type of POE also involves investigation of recent and similar facilities of the same type and compares these with the building being assessed to find out the underlying outcomes of problems. As a result appropriate solutions to the problems are reported and recommended for action.
3. *Diagnostic POE*. Diagnostic POE requires considerable effort and expense to carry out. It is often time consuming and may involve months or years to complete. It utilizes sophisticated data gathering and analysis techniques. The aims of diagnostic POE are twofold. First, it assists in improving existing facilities, and second, the data collected may influence the future design of similar facilities. A combination of various techniques is engaged in carrying out a diagnostic POE including questionnaires, surveys, interviews and physical measurements. Since diagnostic POE requires considerable time to complete, the results are likely to have higher credibility.

22.3.3 Generic POE approach

In order to measure a variety of responses and evaluate a number of issues, several information-gathering techniques are used. These include introductory meetings, touring interviews, scheduled interviews with key personnel or representatives from a population, questionnaires and review meetings. The method used will vary according to the size of the user population and the level of assessment required.

The introductory meeting allows the evaluator to meet with participant groups to explain the process and procedures of the touring interview. This is conducted in small groups of six to twelve people. They are roughly of the same status in an organization to avoid inhibition. During the meeting the evaluator discusses specific questions or problems in relation to a particular group of users and initial complaints may also be identified.

A touring interview may be scheduled with each participant group. The approach involves taking a slow tour through the facility during which each member should be given the opportunity to discuss and express his or her views. Questions about their impression of the areas and opinions concerning functionality may be asked. After the touring interview, a scheduled interview may also be conducted to meet with particular target groups or a representative sample of the user population. The group meeting brings together the users of the facility and provides opportunity for them to address their dissatisfaction and opinions collectively. Such an in-depth interview often effectively identifies information about user satisfaction.

Questionnaires may also be used to conduct a much wider survey to elicit ratings from building users. The advantage of using a questionnaire is that it is a more systematic way of collecting data. By analysing the survey results problem areas can be identified and isolated for further investigation. The survey report can also be a useful tool in presenting a case to senior management. Furthermore, a questionnaire can provide a record for future reference and becomes evidence of work done should complaints arise again in the future. The POE process may end up with a general review meeting. This is the time when essential negotiation takes place and feedback of the evaluation is provided to the participants. During the meeting problems which arose from the previous stages are discussed and agreed recommendations are prioritized for action.

22.3.4 Benefits of POE

Despite its obvious usefulness in understanding current building performance and occupant requirements, POE is an important tool for facility management and for planning of new facilities. Potential benefits of POE range from short term to long term (Barrett, 1995):

1. The short-term benefit of POE is to allow management to have better understanding of the functionality and performance of their building compared with the stated criteria during design. Before improvements can be made, the problem needs to be identified and studied in detail. Active user participation

in the evaluation process plays an important role in this respect. As a result user values are confirmed and respected with regard to the management of an organization, thus stimulating user satisfaction and productivity.

2. Medium-term benefits comprise the use of data collected during the POE process to be a source of knowledge for planning new buildings of similar type. Designers equipped with user feedback on building performance may help to design future buildings that more closely meet the work needs of users.

3. In the long term POE helps to establish databases, generate planning and design criteria for specific building types and enables designers to consider documented past experience. This is important to avoid repeating past errors and to recognize past successes. The accumulated information plays an important part in improving the quality of future buildings and their services to the client. Evaluation results may also improve design practice by making designers aware that their buildings may be the subject of scrutiny. Thus design of future buildings may lead to better value for money to clients and society generally. This concerns not only issues of functionality, but overall sustainability, energy efficiency and environmental impact.

22.4 Case study: an evaluation of a 1960 multi-storey office building

The subject of the POE evaluation was an office building constructed in the 1960s (Kernohan et al., 1992). It had seven storeys and was occupied by eight different tenant organizations. The building had been occupied for over 25 years and had a long history of complaints about its functionality and condition. The evaluation was facilitated by a team of two architects and a representative of the building owner. The objective of the evaluation was troubleshooting and finding solutions to identified problems.

The evaluation procedure consisted of introductory meetings followed by touring interviews and ended with a review meeting to present the survey results. The evaluation was conducted by 10 participant groups of which eight were representatives from various interest groups and the other two were made up of senior managers and maintenance and management personnel. The evaluation began onsite by conducting introductory meetings and touring interviews with the 10 participant groups. During the touring interview photographs were taken in relation to the issues raised.

There were over 100 recommendation statements received. Before the general review meeting, the recommendation statements were categorized under broad headings such as environmental control, entry, workplace, staff facilities and maintenance. Cost estimates were prepared for each category. The general review meeting was attended by all participants. During the meeting participants prioritized the groupings by considering the landlord's attitude to the issue, and the impact of time and budgetary constraints. Participant groups worked together to reword recommendation statements to reflect their priorities and work within the time and budgetary constraints set by the landlord.

The results of this POE were in three broad areas. First, the problems of over-heating in the building and lack of control were overcome by introducing some additional cooling units. However, in the long term they agreed to appoint a consultant engineer to carry out a more detailed analysis of the existing ventilation system. Second, they had decided to modify the existing signage system. New and additional signs would be put up on the exterior of the main entrance and the rest of the public areas. A new directory board was also required in the lift lobby. In addition tenants agreed to ensure their internal signs were consistent and compatible with the others in the building. Finally, since participants complained that the toilets were poorly lit, it was agreed to upgrade the existing lighting system and repaint the walls with a lighter colour.

After the meeting the tenants and the landlord agreed to work together more collaboratively in the future. The building liaison managers agreed to meet on a monthly basis and a representative from the landlord would attend such meetings where appropriate. The evaluation had proved to be successful in identifying problems and bringing people together to solve them.

22.5 Conclusion

POE is now becoming an important area of consideration by building owners, facility managers and construction industry professionals for managing existing facilities and improving future designs. The data obtained from the POE process provides useful information that considerably reduces the uncertainty in making design decisions. POE focuses on evaluation of a building's performance throughout its life cycle and the feedback process is both inevitable and valuable. Consequently, the evaluation is not seen as a linear process but rather a cyclic evolution, which has a goal of continuously feeding information back to improve quality. The accumulated knowledge will prove beneficial to the better utilization of limited natural resources and the improvement of worker productivity. In particular, POE assists in collecting data for future value management workshops and is a significant input source.

One deficiency with traditional POE is the perception that it is a one-off process to confirm that a new building has met the original client requirements. However, more enlightened building owners and facility managers are using the technique more regularly to ensure that buildings continue to deliver appropriate levels of satisfaction to end-users. In some cases this may be combined with wider activities concerned with environmental compliance.

References and bibliography

Baird, G., Gray, J., Isaacs, N., Kernohan, D. and McIndoe, G. (1996). *Building Evaluation Techniques*. McGraw-Hill.

Barrett, P. (1995). *Facilities Management: Towards Best Practice*. Blackwell Science.

Building Science Forum of Australia (1985). 'Post-Occupancy Evaluation: Expectation and Reality', BSFA Seminar, Australia.

Coenen, F.H.J.M., Huitema, D. and O'Toole, L.J. Jr. (eds.) (1998). *Participation and the Quality of*

Environmental Decision Making. Kluwer Academic Publishers.

Ellis, P. (1987a). 'Post-Occupancy Evaluation'. *Facilities*, vol. 5, no. 11, pp. 12–14.

Ellis, P. (1987b). 'Post-Occupancy Evaluation: The Planning Workshop'. *Facilities*, vol. 5, no. 12, pp. 4–6.

Foxall, G. and Hackett, P. (1994). 'Consumer Satisfaction with Birmingham's International Convention Centre'. *Service Industries Journal*, vol. 14, no. 3, pp. 369–80.

Hamer, J.M. (1988). *Facility Management Systems*. Van Nostrand Reinhold.

Kernohan, D., Gray, J., Daish, J. and Joiner, D. (1992). *User Participation in Building Design and Management*. Butterworth Architecture.

Kincaid, D. (1994). 'Measuring Performance in Facility Management'. *Facilities*, vol. 12, no. 6, pp. 17–20.

Maclennan, P. (1991). 'Post-Occupancy Evaluation'. *Facilities*, vol. 9, no. 12, pp. 14–15.

Marmot, A.F. (1991). 'The Good Office: Post-Occupancy Evaluation of Office Buildings'. *Facilities*, vol. 9, no. 12, pp. 10–13.

McCornell, R.L. and Abel, D.C. (1999). *Environmental Issues: Measuring, Analysing and Evaluating*. Prentice Hall.

Preiser, W.F.E. (1989). *Building Evaluation*. Plenum Press.

Preiser, W.F.E. (1995). 'Post-Occupancy Evaluation: How to Make Buildings Work Better'. *Facilities*, vol. 13, no. 11, pp. 19–28.

Prill, M. and Lubimowski, G. (1988). *Evaluation of Four Access Centres in New South Wales*. NSW Department of Technical and Further Education.

Thomas, R. (1999). *Environmental Design: An Introduction for Architects and Engineers* (2nd edition). E. & F.N. Spon.

Thorne, R. and Turnbull, J.A.B. (1991). *Post Occupancy Evaluation Case Study: Technical Report*. Architectural Psychology Research Unit, University of Sydney.

Vischer, J.C. (1989). *Environmental Quality in Offices*. Van Nostrand Reinhold.

23

Environmental
auditing

23.1 Introduction

Environmental auditing is essentially the process of determining whether an organization complies with the regulatory requirements and internal policies and standards that govern its performance. It is an objective and often independent investigation of issues like pollution control, worker health and safety, product safety, energy usage and loss prevention. It has a significant marketing aspect in respect to its ability to position an organization so that its products and services are seen to be more attractive to consumers.

Environmental auditing can be applied to new projects or policies (often referred to in this context as environmental impact assessment), existing facilities, or whole organizations. A subset of environmental auditing that has direct application to the performance of built facilities is energy auditing. The procedures used for auditing environmental performance fit into a larger environmental management system that is used to guide organizational behaviour.

Environmental audit programmes do more than verify environmental compliance. Greeno et al. (1985) indicate that a wide variety of objectives and benefits can be pursued depending on corporate culture, management philosophy, size and market perceptions. Identification of risks, liability exposure and the possibility of greater financial return are three important considerations that govern the implementation of environmental auditing processes.

23.2 The greening of corporate strategies

Ledgerwood et al. (1993) suggest that issues of corporate management are at the heart of environmental problems, as well as the achievement of sustainability goals. They identify twelve principles for integrating an environmental programme into the corporate strategy:

1. Focus on the corporate learning curve. Environmental goals need to be acquired and assimilated alongside financial, operational and personnel goals.
2. From the start, use examples of existing good practice to inject enthusiasm into the environmental programme.
3. Rely on, manage and use technical experts.
4. Gain and sustain top management commitment.
5. Make environmental audit success part of the promotion route for young managers.
6. When possible, convert environmental benefits into financial cost savings and investor attractiveness, and build these into the reporting cycle.
7. Admit mistakes and shortcomings and agree goals with corporate critics for putting environmental problems right.
8. Build environmental auditing into the reporting cycle and use this as a basis for dialogue with insurers, investors and community groups.
9. Set and achieve attainable short and mid-term targets, which can be publicized by key middle and senior corporate managers.
10. Build on environmental successes.
11. Concentrate on discovering points where environmental performance can be converted directly into core financial and quality strategic targets.
12. Respect corporate critics and develop methods of working with their concerns to achieve a 'win-win' outcome.

It is now generally regarded that corporate 'greening' leads to marketplace advantage and increased financial return. For this reason in recent years there has been a significant increase in corporate activity related to environmental performance. Actions can be narrowly applied to issues such as paper usage and recycling potential, or to broader issues such as increasing market share through environmentally-safe design and production processes and packaging strategies. Environmental auditing is the vehicle used to arrive at a greener organization via a systematic and objective manner.

23.3 The auditing process

In some respects, environmental auditing techniques are modelled on traditional financial audits. The basic functions of a financial audit are to verify adherence to standards, to certify that accounting procedures are appropriate for those standards, and to certify the accuracy of corporate financial records. Environmental auditing programmes employ verification techniques for confirming compliance with federal, state and local regulations, assessing the adequacy of environmental management programmes to ensure policies and procedures are followed and standards are adhered to, and verifying the validity of environmental records and reports.

The audit process essentially comprises three stages: pre-audit activities, on-site audit, and post-audit follow-up. Pre-audit activities include defining the scope of the investigation, team selection, development of the audit plan and allocating necessary resources. Post-audit follow-up concerns implementing change to rectify or improve existing practices. The on-site audit can be categorized as comprising five key steps:

1. *Understanding internal management systems and procedures*. The auditor's understanding is usually gathered from multiple sources such as staff discussions, questionnaires, facility tours and, in some cases, a limited amount of verification testing.
2. *Assessing strengths and weaknesses*. The auditor must assess the strengths and weaknesses of existing systems and procedures. Quality control records such as clearly defined responsibilities, authorization, the awareness and capabilities of staff, documentation and internal verification are used to establish performance indicators.
3. *Gathering audit evidence*. Gathering evidence of actual performance is the basis upon which the auditor determines compliance and forms opinions. Audit evidence can be collected through inquiry, observation and testing.
4. *Evaluating audit findings*. Gathered evidence needs to be assimilated and evaluated. This process is achieved through meetings and discussions with key personnel.
5. *Reporting audit findings*. A formal report concludes the on-site audit and is intended for management. It is therefore a summary of key findings together with recommendations for improvement.

The auditing process is quite flexible and will be adapted to suit the particular circumstances of the task. The use of an audit team independent of the organization has considerable merit and increases the likelihood of objective assessment.

23.4 Energy auditing

Energy auditing is a systematic way of gathering and evaluating information with regard to the quantity and type of energy used, and is a specialized form of environmental auditing. It comprises the periodic survey, measurement, reporting, analysis and examination of an energy system for individual plants, a production process or an entire organization.

Energy auditing is defined as 'a periodic examination of an energy system (or part of the system) to ensure the most appropriate sources of energy are employed, and this energy is used as efficiently as possible' (Department of Primary Industries and Energy, 1994a, p. 1). The goals of energy auditing are to promote energy efficiencies, to identify areas of potential savings in energy and its related expenditure, and to promote energy management so to achieve, maintain and recognize further potential savings.

Energy auditing forms part of the effective control of energy use and its related cost within an organization. The benefits of energy auditing are not just reducing operational costs but also conserving the environment. As a result of better planning and close monitoring of energy use, there will be reductions in energy consumption that may lead to conservation of limited natural resources, and more importantly, harmful emissions of carbon dioxide due to combustion of fossil fuels can be reduced.

23.5 The energy auditing process

Energy auditing plays an essential role in energy management programmes. It enables organizations to understand more about the way they use energy resources. For example, in promoting energy auditing and responding to the national greenhouse challenge, the Australian Government in 1991 announced the Enterprise Energy Audit (EEA) Programme. The main objective of the programme is to encourage enterprises to identify cost-effective energy savings and hence reduce greenhouse gas emissions. At the same time the programme provides a database of energy use standards which represents the best practice within the sector in terms of both energy efficiency and emissions control. The feedback from energy auditing is important for further research and development in energy-efficient technology. The programme encourages organizations to undertake energy auditing by offering monetary assistance up to $5000 per audit or 50% of the cost of an energy audit (whichever is the lesser amount).

Energy auditing comes in various forms that are flexible enough to suit different organizational requirements. Energy auditing can be as simple as checking monthly or quarterly energy bills. This does not involve a detailed investigation and can be performed with little specialist knowledge about energy. An authorized auditor may not be required in this respect. Alternatively, basic energy auditing can be upgraded by checking and keeping site records of main meter and sub-meter readings. Nevertheless, some engineering knowledge is required for checking the reliability of site data.

A full energy audit may be conducted by undertaking a site survey to obtain information like flow, control and management of energy use. It helps to identify and evaluate opportunities for savings by examining energy use breakdown in terms of fuel types and end-users. The proportion of energy consumption by heating, lighting and ventilation is measured, recorded and analysed. Energy auditing of this type may be carried out for financial purposes. It provides management information to assist in decision-making in determining the priorities for a more detailed investigation, investing in energy efficiency measures, raising energy awareness, identifying cases for direct action and deciding future fuel budgets. It is normally based on the financial year of the organization.

Energy auditing of this type is conducted by an energy auditor (which may be a person, a team or an organization). The auditor is responsible for co-ordinating the provision of all meters, instruments and equipment necessary to meet the intention of auditing, determining their accuracy, preparing a report and making recommendations for further improvement. The technical procedures and reporting formats of energy auditing are governed by standards and codes of practice.

23.6 Energy auditing procedures

No matter what form energy auditing adopts, the usual procedures comprise survey and measurement of energy use, data analysis, evaluation and implementation. They are therefore not dissimilar to generic environmental auditing procedures.

Data collection is the initial step in carrying out energy auditing. The required information includes the description of the site, nature of operations on site and the type of energy use. Periodic records of energy consumption and cost in the form of fuel invoices and accounts from suppliers are useful data. They may be presented monthly or quarterly depending on the size of an organization. Invoices may be based on actual meter readings or estimates. Incomplete or estimated invoices may be supplemented by site records of main metered or sub-metered energy consumption readings and stock levels. Thorough checking of the collected information is essential to ensure that a true and complete record is provided. A set of as-built drawings is a useful tool in providing information like the schematic layout of the operations on site and locations of various items of plants and fuel types. With fuel invoices, site meter records and sets of drawings in hand, energy auditing is moved on to the next step of data analysis.

Data analysis concerns analysing total energy consumption and expenditure from invoices and meter readings to reflect the relative importance and efficiency of a system. The total energy consumption of each fuel type is listed and converted to gigajoules (GJ) based on the information from invoices counter-checked by site meter records. The total energy consumption and expenditure can be tabulated or recorded by computer spreadsheets on a daily, monthly or quarterly basis. These formats enable calculations to be made and present information like fuel types, units of each fuel, conversion factors (individual unit converted to GJ), total energy consumption, percentage of total energy input, total cost and percentage of total annual cost.

The recorded fuel consumption is then depicted using pie charts by fuel types or end-users with percentages attributed to them respectively. Line and bar graphs are also used to show movements in fuel consumption. Climatic records are employed to compare against graphs in identifying correlation with energy consumption trends. The pattern of energy use can be examined monthly to indicate general trends or seasonal patterns of energy use. The seasonal/trend analysis indicates seasonal patterns of use and base loads. On the other hand, seasonal or cyclical patterns indicate major seasonal loads whilst general upward or downward trends reflect changes in load or efficiency. A lack of a clear pattern may suggest a lack of control and obvious anomalies can often be explained by errors in the original data. Where records are available for several years, the information may be plotted to establish historical trends in consumption, expenditure and activity levels. Hence potential energy savings can be worked out by comparing the fluctuation in the graphs and comparing energy consumption levels.

Apart from analysis of energy consumption by fuel types, energy consumption can also be analysed in the form of end-users, main services, different sections of a single building or different buildings in a multiple site. A list can be drawn up to identify areas of unacceptably high energy consumption. This process isolates major energy-using areas for a further detailed site survey and investigation. Corrective measures may be required like replacing dilapidated plant or equipment, changing occupancy patterns (e.g. hours of operation) and increasing awareness of energy conservation among operatives in order to bring back efficiency to acceptable levels.

The analysis of yearly fuel consumption and expenditure can act as an energy performance indicator. It becomes a foundation for future energy use monitoring and may be used to detect energy waste and inefficiency. The performance indicator can be projected to forecast future energy consumption and set fuel budgets. It provides a basis to evaluate the future energy performance of an organization and indicate the areas requiring closer examination. It can also be used to compare energy use with buildings of similar type or with a preset standard or target to reflect efficiency. Energy saving may be based on an estimated percentage reduction in annual consumption or a direct calculation from a reduction in the annual fuel cost, load, operating hours or energy loss.

Technical and financial evaluations of energy consumption are two important aspects in considering options for savings. Technical evaluation ensures that implementation is physically possible within the site and considers requirements for design, installation, supervision and future maintenance. Financial evaluation involves evaluating the return available from investment in corrective actions and to ensure that a positive return is available within the expected life of the measure. It compares the returns of alternative measures or other opportunities for investment, and hence determines an order of priority for implementation. There are different methods of evaluating return and the common techniques in this respect are payback period and discounted cash flow. The former is generally for measuring returns within five years or involving only very minor investment, while the latter is used to evaluate large projects with long-term impacts.

Based on information from technical and financial evaluations, recommendations may be categorized into no/low, medium and high cost measures. They are presented in order of priority for implementation. The order of implementation can be prioritized by measuring the net present value or alternatively the internal rate of return. Measures with a return higher than the basic profitability of the business should offer a higher opportunity for implementation.

23.7 Implementing recommendations

The principal objective of energy auditing is to identify worthwhile opportunities for saving energy and its associated expenditure. Corrective measures will be selected from a list of recommendations. The selection process considers the availability of funding and management time for implementation. Priorities will be given to those measures with competitive financial commitment which are within the capacity of management to implement and support. After selecting a measure, an action plan is produced to identify the tasks involved in implementation. The implemented measures should be monitored closely and checked regularly against the preset target.

Implementation of auditing recommendations involves setting targets for improving existing energy systems or installing new energy systems to achieve the preset targets. Normally recommendations will start with improving performance of existing energy systems through better control. This may involve the installation of building management systems (BMS) and controls. Expertise may be required in attendance to monitor the BMS so closer control can be achieved of existing temperature and time settings. If improving performance of existing energy systems cannot fulfil the requirement, new systems may be required to supplement the existing energy systems. Passive systems may be introduced by altering the building form and fabric without resort to mechanical devices. The main aim of passive systems is to control heat gain and heat loss during different seasons of the year. However, the existing building structure may restrict the application of this system. Life-cost studies may be required to assist decision-makers to determine whether it is beneficial to adjust the existing structure to accommodate passive systems.

Other systems like thermal storage, heat exchangers, solar collectors and the like may also be useful to store energy during off-peak hours, collect solar energy during the daytime and reuse waste air to preheat fresh air entering into the building. Alternatively, saving energy costs may be achieved by shifting from high energy cost sources to low energy cost sources such as from expensive electricity to cheaper natural gas. Changing patterns of work may also help. Recent research has noted that people travelling to work consume a great deal of energy daily. With the development of modern telecommunication technology, people may be able to work from home to reduce energy consumption and the pressure on transportation systems.

23.8 Benefits of energy auditing

The primary aim of energy auditing is to understand the current energy system and to identify measures for energy saving. Energy saving is crucial to the environment in several aspects. Saving energy in a plant, a process or an organization will generate less carbon dioxide and other emissions in the environment that enhance global temperatures. Furthermore, it helps to reduce the pressure on the demand of finite natural energy resources, particularly fossil-based fuels, thus assisting with their conservation. Energy auditing provides useful information to enable use of energy in a more efficient manner.

Energy auditing reveals energy performance of a plant, a process and an organization that can be used to compare with others of similar nature. The energy performance indicators can be useful in establishing priorities for action and can detect anomalies for immediate action or detailed investigation. It helps to determine investment in energy efficiency measures. The breakdown of energy cost by fuel types and end-users helps to establish energy management systems to monitor energy consumption in the long run. Estimates can also be projected for future energy consumption and budgets can be set for fuel purchases based on current auditing information.

Energy auditing can lead to improvements in working conditions. It may recommend installation of insulation to the building fabric or draught proofing to minimize heat loss in the winter and heat gain in the summer. The increased comfort to working environments may lead to increased productivity that may in turn lead to further financial benefits.

The direct benefit of energy auditing is cost saving. Cost saving can be achieved through less energy consumption as a result of improved efficiency and better control. The comparison of normalized performance indicators with buildings of similar type can then indicate whether there is likely to be a good opportunity for improvement and help set priorities for immediate action.

23.9 Future importance

Environmental auditing, and in particular energy auditing, may take on greater importance in the near future if regulatory control is increased as a response to deteriorating environmental resources. For example, new and existing buildings may need to conform to maximum energy limits per annum, and in order to determine compliance some form of independent audit may be required. Environmental protection agencies may also require objective assessment of pollution control and waste disposal practices. Exposure to financial penalty or litigation will encourage organizations to raise not only their performance but also the quality of the systems that govern this performance.

The efficient utilization of energy is becoming an important issue to tackle environmental problems and is at the forefront of environmental performance for built facilities. Energy efficiency brings many benefits including saving energy cost, saving non-renewable fossil-based fuels and limiting effects to the environment through reduced greenhouse gas emissions. Energy auditing is a systematic way of understanding current energy systems by identifying opportunities for cost and energy savings. It establishes a base for both patterns and quantities of energy use from which subsequent comparisons can be made and anomalies can be detected. The data collected and analysed can be further projected to forecast future energy demand and set energy budgets. Energy auditing is a useful tool for management in understanding current energy systems and patterns of fuel used. The more knowledge acquired the better the future control and monitoring of energy usage will be. If properly and continuously performed, energy auditing can lead to changes that reduce energy use, increase employee comfort and ultimately help to improve environmental quality.

23.10 Conclusion

Environmental auditing is an important component of asset management in that it establishes compliance with internal and external codes of practice and identifies areas of deficiency. In an increasingly litigious society it is critical that organizations protect themselves from the risk of financial loss as a result of their business activities. This may arise from the methods employed to deal with waste and limit pollution, to the energy and products consumed in building operation. Although the technique is still evolving, it is clear that environmental auditing is an area of considerable interest to organizations worldwide.

References and bibliography

Chartered Institution of Building Services Engineers (1991). *Energy Audits and Surveys Application Manual*. Unwin Brothers Ltd.

Department of Primary Industries and Energy (1994a). *The Energy Audit*. Energy Management Advisory Booklet, AGPS, Canberra.

Department of Primary Industries and Energy (1994b). *Financial Evaluation of Energy Management*

Measures. Energy Management Advisory Booklet, AGPS, Canberra.

Editor (1991a). 'Enterprise Energy Audit Program Begins'. *Australian Energy Management News*, no. 13, May, p. 6.

Editor (1991b). 'Energy Audits Highlight Large Savings Potential'. *Australian Energy Management News*, no. 17, August, p. 19.

Elliott, D. (1997). *Energy, Society and Environment: Technology for a Sustainable Future*. Routledge.

Ellis, P. (1987). 'People and Organisations: The Hidden Energy Resource'. *Facilities*, vol. 5, no. 2, February, pp. 7–9.

Greeno, J.L., Hedstrom, G.S. and DiBerto, M. (1985). *Environmental Auditing: Fundamentals and Techniques*. John Wiley & Sons.

Griffiths, A. (1992). *National Energy Management*. AGPS, Canberra.

Herzog, P. (1997). *Energy-Efficient Operation of Commercial Buildings: Redefining the Energy Manager's Job*. McGraw-Hill.

Hodgson, P.E. (1997). *Energy and Environment*. Bowerdean Publishing.

Howard, N. and Roberts, P. (1995). 'Environmental Comparisons'. *The Architects' Journal*, September, pp. 46–47.

Hutchinson, A. and Hutchinson, F. (1997). *Environmental Business Management: Sustainable Development in the New Millennium*. McGraw-Hill.

Ledgerwood, G., Street, E. and Therivel, R. (1993). *The Environmental Audit and Business Strategy: A Total Quality Approach*. Pitman Publishing.

Moss, K.J. (1997). *Energy Management and Operating Costs in Buildings*. E. & F.N. Spon.

Osie, G. (1991). 'Energy Audits get the Green Light'. *Facilities Design and Management*, June, pp. 62–63.

Ristinen, R.A. and Kraushaar, J.J. (1999). *Energy and the Environment*. John Wiley & Sons.

Royal Australian Institute of Architects (1983). 'Energy Auditing'. *Practice Note PN 78*, RAIA.

Tuluca, A. (1997). *Energy Efficient Design and Construction for Commercial Buildings*. McGraw-Hill.

Williams, B. (1985). 'The Energy Audit'. *Facilities*, vol. 3, no. 4, April, pp. 3–6.

24

Facility management

24.1 Introduction

Prior to the emergence of facility (or facilities) management (FM), activities like property management, maintenance, operational performance, leasing, selling, buying and planning were carried out by somewhat isolated teams of building maintenance staff, office managers, property portfolio managers, property managers, accountants and financial officers. No one individual co-ordinated all the players. FM is the emerging discipline that has replaced fragmented responsibility and decision-making with centralized management.

FM has always been practised to some extent by organizations, yet only recently it has emerged as a discipline. The International Facilities Management Association (IFMA) (cited in Hamer, 1988, p. 1) define it as 'the practice of coordinating the people and the work of an organisation into the physical workplace'. FM is about the alignment of an organization's facilities with its strategic plan. This focus stresses holistic and integrated strategic planning, which has been made possible by advances in information technology (IT) and made significant by the emergence of the information age and the need for ecologically sustainable development (ESD).

The proliferation of information and data created by computers, mass media and telecommunications is completely revolutionizing the economy, society and the way in which work is being carried out. Globalization and rapidly evolving products and service markets has created a need for flexible organizational structures and adaptable facilities to service those organizations. FM is a synthesizing discipline that has emerged in response to the need to manage change.

24.2 Computer-aided facility management

Traditionally FM was conducted using file cabinets, file folders, as-build drawings, architectural floor plans, file cards and manufacturers' catalogues. These techniques of FM have proven too labour intensive, inaccurate, inconvenient and time consuming. Nowadays with the help of information technology, FM has been made more efficient. FM centres on effective collecting, recording and interpreting of data electronically. IT has enabled the facility manager to readily collect data at a single point from previously autonomous parts of the organization and aggregate these sources into statistical reports which senior management can use for decision-making. IT also enables data produced in facility design and planning to become the basis of ongoing management decisions, thereby breaking down the traditional boundaries of design, construction, project management and maintenance into a continuous cycle.

IT used for FM has become known as computer-aided facility management (CAFM) which has the following benefits:

1. *Increasing productivity and saving costs*. Labour productivity is closely linked with cost savings. As people work faster, less time is spent on production and therefore the unit cost of production is reduced. Labour productivity is greatly affected by the nature of work and the availability of facilities. FM is labour intensive because it requires collecting, interpreting and processing a great deal of information. With the assistance of computers, the overall time of professional, technical and clerical staff is reduced. Hence less time is required to complete the work and thus the cost attributable to the labour involvement is reduced.
2. *Enhancing design quality*. Computers enable work to be completed with greater speed and more accuracy. Therefore, the facility manager is able to consider many more alternatives quicker and easier and is more likely to identify options that represent the best value for money.
3. *Reducing error*. Minimizing error means minimizing loss. The daily operation of a business is complicated and errors are inevitable. Nevertheless computers can help to minimize human error especially for those operations of a repetitive nature. Financial and operational management of an organization involves making a large number of individual decisions, co-ordinating the work of many different people and departments and producing a great deal of highly detailed information.
4. *Enhancing management effectiveness*. Computers are practical in the monitoring of resource expenditures and achieve greater budget and scheduling predictability by replacing relatively unpredictable human performance with more reliable machine performance. It also enables tightening control over information access and security.

IT within the facilities themselves has also led to the demand for FM. IT including networked office computers, telecommunications, building management and security system installations are likely to change faster than the building fabric. The facility manager is needed to organize installation and maintenance including systems and manage the relationship between them and the building fabric.

Another aspect contributing to the growth of FM is the increasing emphasis on ESD resulting from more stringent environmental statutes and regulations. The ESD imperative will inevitably lead to greater reuse and adaptation, rather than development and new construction, and the facility manager will be necessary to co-ordinate the task.

24.3 Facility management models

Different organizations have various models for FM. Generally FM models are flexible enough to satisfy a range of needs, and therefore its practice can vary greatly. Barrett (1995) suggests five models for FM as currently practiced in the market:

1. *Office manager model*. Some organizations do not have separate departments to undertake facilities work. It is because they may be either too small to warrant one or they do not own the building they occupy. Therefore, they have no real control over the facilities work and are unwilling to commit their time and money into such considerations. FM is generally conducted by office managers as part of their general duties. Other specialist work as required is carried out by external consultants or contractors through the administration of service contracts and leases.
2. *Single site model*. This model is practised in large organizations located on a single site. These organizations usually own the buildings they occupy and therefore are willing to commit more time and money to their proper management. In this model FM will be undertaken by a separate department with a full-time facility manager with broad responsibilities. However, external consultants and contractors may still be required to carry out specialist work.
3. *Localized site model*. This model applies to even larger organizations with a headquarters and several branch offices located in the same metropolitan area. In this model a facility department in the organizational headquarters provides facilities policies, budgetary control and technical assistance, while facility departments in branch offices are responsible for undertaking daily activities related to their own buildings.
4. *Multiple site model*. This model is similar to the localized site model except the branch offices are located nationwide. The function of facility departments in the headquarters and branch offices is the same as the localized model, but very large organizations may require regional facility departments to co-ordinate activities within the same region.
5. *International model*. This model applies to large international organizations with branch offices located in various countries. The function of facility departments operates on a much wider and higher level. Usually local facility departments are required to overcome the problems of different languages, currency and legislation systems.

24.4 The strategic facility plan

FM is a very broad field and activities include property management, facility planning and operations, maintenance and facility support services. However it requires more than the summation of these operational activities, in that these functional tasks must be integrated and form the basis of holistic decision-making.

The facility manager uses information on the operation and performance of numerous dynamic parts of the organization's facilities, to formulate a Strategic Facility Plan (SFP). An SFP is a long-term holistic life-cycle view of facilities in accordance with an organization's strategic objectives.

Any strategic plan must be a flexible 'flight plan' adjusted to changes in its environment, the intention of members of an organization, or the availability of resources. To develop and derive such a dynamic, flexible and responsive SFP, a facility manager must understand the corporate strategic planning process.

Strategic planning provides a framework for decision-making throughout an organization. It prevents off-the-cuff decisions, encourages and permits the testing of value judgements, stimulates the effective discharging of responsibilities, and is a valuable communication device.

24.5 Strategic planning

Strategic planning is the process of positioning an organization so that it can prosper in the future. It deals with the futurity of present decisions and long range connotations, the obsolescence of existing products or processes and the new ones that take their place. The following definitions apply to strategic planning activities:

1. *Mission statement*: A statement of what an organization is, why it exists and the unique contribution it can make.
2. *Objective*: A desired or needed result to be achieved in the long-run future.
3. *Strategy*: A set of decision rules and guidelines to assist orderly progression toward an organization's objectives.
4. *Goal*: A specific time-based point of measurement that the organization intends to meet in the pursuit of its broad objectives.
5. *Programme*: A time-phased action sequence used to guide and co-ordinate operations in the pursuit of a goal.
6. *Project*: The implementation of a programme or part of a programme by identifiable activities and resources and leading toward the attainment of specific goals.

A strategic plan would normally contain the following components:

1. Executive summary.
2. Description of the organization.
3. Objectives and goals.
4. Environmental considerations.
5. Internal assessment.
6. Strategy.
7. Evaluation of alternatives.
8. Tactics.

9. Economic projections and results.
10. Monitoring.
11. Contingency plans.
12. Summary and conclusions.

Some might say that the essential objective in any business continues to be profit, as no company can improve the environment, society and give its employees security if it is bankrupt. If a company were to maximize customer service it would probably impair profit. However, others might say each organization has a spectrum of objectives, including social responsibility, which to some extent may be compatible with profit maximization.

Objectives are sometimes ordered into a pyramid, with the corporate mission at the peak, a group of objectives in the next wider step below, followed by the corporate goals at the base of the pyramid.

The mission statement exists as the highest guiding principle within the organization and is usually simply expressed in a paragraph or two. Objectives are more specific than the mission statement and express desired results to be achieved in the long-range future. A set of objectives are challenging, attainable and mutually consistent aims that can be used to guide all parts of an organization in its development. Goals are specific accomplishments that the organization intends to achieve by a specific date and are stepping stones in the pursuit of the organization's broad objectives.

24.6 Stakeholders

Each business has many claimants. For the most part the claimants do not have the authority to impose objectives on the company. Objectives are written by all levels of the company's management. When objectives relating to particular stakeholders are included it is because explicit statement of such objectives will be beneficial to the company over its strategic planning horizon. Stakeholders commonly include suppliers, competitors, special interest groups, government, stockholders, executives, bankers, unions, employees, consultants, contractors and customers.

24.7 Environmental/contextual factors

Strategic planning is dependent upon the idea that an organization must balance its internal capabilities with its external environment or context. Since this balance is achieved by apportioning the organization's resources to achieve goals consistent with external characteristics and demands, the context of strategic planning is extremely important. The borderline between external and internal factors is not clear cut. Generally speaking the environmental factors are those which are not directly controlled or influenced by the organization.

Forecasting changes in the environment of an organization is extremely difficult because of the sheer number factors over which the organization has little or no control, the range of both quantitative and qualitative factors involved and the uncertainty as to which of the many exogenous factors will become important in the future. However a decision must be made as to which of all possible factors are relevant to the organization's costs, operations and profitability. Sometimes, only a few factors in the environment turn out to be highly significant.

24.8 Determining the strategy

'Strategy' is not the same thing as 'strategic planning'. The strategy is the general or macro approach to be followed in achieving an objective. Strategic planning follows the guidelines of the strategy statements but adds the details of dollars, volumes and timing, and the tactical and operation planning of lower organizational levels. As objectives and goals may be to some extent inconsistent, priorities have to be set as to what will be undertaken and what may have to be neglected. This choice is outlined in the strategy.

24.9 Alternatives analysis

Methods of generating alternatives include examining existing business areas to see what could be either reduced, maintained or increased or by undertaking a SWOT analysis, asking of each alternative:

1. Does it capitalize on company strengths?
2. Does it eliminate or reduce company weaknesses?
3. Does it allow the exploitation of opportunities?
4. Does it reduce potential severe threats?

A choice must be to establish that the set of alternatives includes all potentially successful means of attaining a particular objective. Once alternative means for accomplishing objectives have been analysed the next step is to quantify and set targets or goals which then become part of a company's strategic objectives.

Ethos is the disposition, character or attitude peculiar to a specific people, culture or group that distinguishes it from other people or groups. Ethos refers to the manner in which an organization behaves towards its employees and those external to the organization. What will not do on moral grounds presents a constraint that impacts upon strategic choice. Questions should be asked of each alternative:

1. Does the alternative under consideration call for morally objectionable actions?
2. Are any ethological constraints infringed?

Once a number of alternative strategies have been ruled out, the remaining list may be more closely scrutinized. Every alternative should be looked at from two perspectives: operational objectives and project objectives. The operational objectives consist of proposals for modifying the existing business, whereas the project objectives involve proposals for entirely new activities.

Quantitative models are extremely valuable in this process. Computerized sensitivity testing is often useful in evaluating alternatives and estimating the effect on the return on investment of a particular alternative. Quantitative models commonly used include cash flow models, income models, production models, purchasing models and life-cost studies.

24.10 Contingency planning

A contingency is something that may happen but is not certain to happen. All environmental factors have a strong contingency aspect. To a large extent contingencies are accounted for by the strategic planning process itself, as plans are modified each year for the entire planning horizon, and by alternatives analysis, as a particular contingency is often contained in one of the alternatives. Sensitivity analysis carried out using quantitative computer modelling enables the analysis of the impacts of a range of contingencies.

24.11 Feedback

Having decided upon the alternatives, strategic planners should ask other members of the organization:

1. Can the proposed strategies realistically be executed?
2. Does the alternative give the company a performance risk curve compatible with that of its shareholders?
3. Does the company have the competence to carry it out?

Sometimes feedback requires changes to be made to the objectives or mission statement and the whole process must start again. If however all agree that the objective is acceptable, at this point each function or department derives its own mission statement and starts the strategic planning process for that function.

24.12 Functional plans

Functional plans are prepared by each division or department, examined at the corporate level and fed back as necessary for changes or modifications. Plans from all organizational groups must be co-ordinated and if necessary subordinated into the overall strategic plan.

A functional plan is a strategic plan that has a tactical component to cover short-term activities such as the supervision of operations and the co-ordination of current functions.

FM has in the past been seen to involve tactical planning only, as property portfolio managers picked up the responsibility for strategic planning. In some organizations operational matters are separated from property portfolio management by a broad chasm in the management structure. This has resulted in many problems including the management of facilities in a primarily short-term manner.

The SFP is a functional plan that has both tactical (operational and short range) and strategic (long range) components, all of which must relate back to the overall corporate strategic planning process. Tactical planning should be a subset of strategic planning but include more operational details. These activities may be carried out in-house or the decision may be made to out-source them to external service providers. The decision to out-source is not a forgone conclusion and requires informed judgement based upon a complex set of issues.

The functional FM activities include space, financial, maintenance, operational services and assets management. These are described separately in more detail.

24.12.1 Space management

It is a strategy for all business to control expenditure and generate maximum profit. This ultimate target can only be accomplished if premises are operating at peak efficiency. Space planning is, therefore, an important activity in FM to help achieve this goal. It is a technique that analyses and modifies the physical occupation of the building into the best arrangement. It allows smooth interaction of activities within a planned space. It is not just satisfying the space needs of all employees but also improving ease of communication, material and information flow, and minimizing fixed operating expenses and redundant activities.

Space planning consists of developing standardized units of space like workstations, storage, equipment and circulation that accommodate various functions and supportive activities. Standardized units of space are derived from a detailed study of the organization's experience, individuals' daily working patterns and departmental consultation. As a result unit spaces are established and are put together to achieve the appropriate layout to suit various types of activity in the organization. Space planning aims at utilizing the existing space in the most efficient way and maintaining work environments to ensure efficiency and productivity.

24.12.2 Financial management

Financial management is another important activity of FM. It is an important task of the facility manager to identify the amount of money that the organization has allocated to run the premises. Indeed, financial management is all about setting the budget, controlling the budgetary cost and maintaining the database of historical cost data for future references. Budgeting is useful to control expenditure and plan for future projects. One of the responsibilities of the facility manager is to prepare a capital budget plan outlining expenditures for new construction, major repairs or renovations, and replacing dilapidated plant and equipment annually. Once the budget plan has been established, the facility manager has to monitor and control all cost items and expenditure from initial estimates to reconciled invoices throughout the life and budgetary cycle. The objective is to ensure that the planned budgets are properly managed and that all opportunities for improvement and cost saving are investigated.

24.12.3 Maintenance management

Maintenance is the work required to preserve or restore buildings, plant and equipment to their original condition and, where possible, to increase their life expectancy and reduce energy consumption. The main concern of maintenance is the building fabric, external works, infrastructure and plant and equipment. The work includes periodic servicing, preventive maintenance and scheduling of routine services such as repair work and replacement. It helps to avoid the risk of loss of amenity due to performance reduction or unexpected failure.

Computer databases are very useful in this respect. They help to record information like location, position, name, function, rating, condition, anticipated life, date of last inspection, next action date and service date of plant and equipment. The facility manager has to check, update the database regularly and identify appropriate remedial actions.

Maintenance can be undertaken either by in-house personnel or by employment of external contractors. For larger and more complex organizations, maintenance departments may be required to ensure continuity of staff. For smaller and less complex organizations, the routine maintenance work may be better carried out by external contractors on a needs basis. This is more flexible in nature and has no involvement of personnel management other than supervision. No matter whether in-house maintenance or contracting is adopted, it is the responsibility of the facility manager to allocate resources, ensure efficient service and monitor quality at all times.

24.12.4 Operational services management

Operational services are regarded as supportive activities that serve occupants in an organization. Operational services management helps to provide a normal environment for the business to function. The provision of operational services enables business to operate efficiently and smoothly within the work environment. It includes providing security and safety for the premises, welfare facilities, voice and communication, crèches and cleaning services. The extent of operational services is governed by the size of an organization, type of workforce, working pattern and type of business.

Security and safety are important issues in providing an efficient working environment for staff. This can be as simple as providing reception services. A reception counter located at the entrance of a building often serves to control people in and out and restrict casual access. For large or more complex organizations, especially those conducting sensitive activities, more sophisticated security systems may be necessary. This may consist of uniformed security guards patrolling the premises day and night and security systems like close-circuit television, staff identification cards, digital locks and card entry access. The facility manager must work together with building designers to establish a security plan to monitor the operation of all physical systems in use and address breakdown or breach. The things that are normally looked at are all possible points of entry into the building, identifying areas of higher sensitivity, anticipating the type of attack and establishing security procedures to integrate staff function and circulation.

Staff welfare facilities are another aspect of operational services that fall within the province of the facility manager. Staff welfare includes providing catering and sport facilities within or outside the premises. The reward for providing staff welfare is higher efficiency and improved productivity. The organization, consequently, builds up a reputation of looking after staff, which engenders loyalty and attracts quality individuals to work for the business.

The extent of staff welfare facilities depends on the size of the organization. In small organizations staff welfare may just be confined to providing tea and coffee-making facilities. Some organizations may be large enough to provide catering and sport facilities to staff. The facility manager, in such circumstances, needs to set up catering services and provide the required space within the building. If an organization engages catering staff direct, the facility manager may be required to manage their work and arrange for any required equipment. If the catering services are contracted out to a catering company, the facility manager may be required to draw up detailed specifications of services, procedures for tendering and contracts for obtaining a suitable catering specialist. Likewise, sport facilities may cover the provision of exercise rooms, swimming pools and sport clubs. It is the responsibility of the facility manager to plan, manage and control the cost of sport facilities within budget constraints. The facility manager may have to review and revise the provision of staff welfare facilities annually and assess the value for money of such amenities.

Routine cleaning is part of general maintenance but occasional cleaning is required to deal with specific or emergency situations like decontamination, spillage of chemical substances or vandalism. They directly affect the usual operation of business and can be hazardous to the workforce. Some large organizations even provide crèche facilities to encourage female staff back to work after their maternity leave. Fire safety procedures and drills are further examples of operational services management.

24.12.5 Assets management

Assets management usually refers to logged inventories that are normally used for depreciation calculations, capital allowance assessments and basic lists of the assets of the business. Inventories require considerable amounts of effort to create and demand regular review and updating in order to maintain their accuracy over time. Assets management also involves the preparation of asset registers, asset tracking, furniture inventory, stock management and control.

Asset registers are established to identify all assets by their make, model, serial number, cost, age and location. Computer databases are an ideal tool in this respect because they are flexible to use and allow information to be updated and retrieved in a quick and easy manner. The asset registers must be checked and updated constantly to maintain their accuracy, and once established, they become a working document for the facility manager in the daily management of inventory. The facility manager may need to check the assets against the registers regularly and carry out appropriate corrective actions. The facility manager may also be required to track down incomplete, missing, mis-positioned or damaged equipment. At the same time it may be necessary to review equipment performance to ensure that machinery is not overworked and its service quality is maintained. The asset registers can also be useful for annual audit and insurance cover purposes.

24.13 Measurement and performance standards

The meshing of tactical and strategic planning is difficult if they do not share the same units of measure. It is crucial therefore, particularly when using information systems, that standard units of measurement are developed that can be applied to both short-term tactical and long-term strategic plans.

Performance standards should be set before action is taken, performance measured against these standards and variances between standards and results evaluated. Interpretation and evaluation of these results should flow through the organization. As such the feedback mechanism is a function of the organizational structure and the intangible factors of organizational behaviour.

In a business organization, results are ultimately expressed as a return on resources employed. Standards, measurements, variances and improvement proposals are also expressed in financial terms. Where strategic plans are taken seriously they are firmly interlocked with company budgets. The process of financial reporting and measurement against preset budgets interlocked with long-term strategic plans is so useful that it justifies substantial resourcing and effort. Effective financial control requires that detailed information be generated at each major operating centre and aggregated for the consideration of upper management.

24.14 Conclusion

FM anticipates change and assesses its impact upon the effective use of built assets. There are many changes in the market that need to be detected and interpreted, such as economic and business cycles, growth and recession. Today successful businesses must anticipate and adapt their working patterns in response to ever changing opportunities in the market. Therefore, management decisions have to be made at the right time. FM enables the organization to identify business and facility life cycles and be able to relate them to avoid redundancy and obsolescence.

The challenges for FM are keeping pace with technology, empowering people in organizations so that they are at their most effective, organizing the service to meet business and user needs and promoting corporate identity.

But perhaps the most exciting challenge for FM is its ability to effectively manage the growing stock of built assets in the context of increasing environmental concern over resource usage and energy consumption. If ecologically sustainable development is to be approached, then surely FM must have a vital role to play.

References and bibliography

Alexander, K. (1992). 'Facilities Management in the New Organisation'. *Facilities*, vol. 10, no. 1, pp. 6–9.

Alexander, K. (1998). 'Facilities Management: A Strategic Framework'. In *Facilities Management: Theory and Practice* (K. Alexander, ed.), E. & F.N. Spon, pp. 2–11.

Atkin, B. and Brooks, A. (2000). *Total Facilities Management*. Blackwell Science.

Barrett, P. (1995). *Facilities Management: Towards Best Practice*. Blackwell Science.

Bernard Williams Associates (1999). *Facilities Economics*. Building Economics Bureau Limited.

Cotts, D.G. (1999). *The Facility Management Handbook* (2nd edition). AMACOM.

Goumain, P. (1989). *High-Technology Workplaces*. Van Nostrand Reinhold.

Hamer, J.M. (1988). *Facility Management Systems*. Van Nostrand Reinhold.

Kaiser, H.H (1989), *The Facilities Manager's Reference: Management, Planning, Building Audits, Estimating*. R.S. Means Co.

Leaman, A. (1992). 'Is Facilities Management a Profession?'. *Facilities*, vol. 10, no. 10, pp. 19–20.

McGregor, W. and Then, D. (1999). *Facilities Management and the Business of Space*. Arnold.

Park, A. (1994). *Facilities Management: An Explanation*. Macmillan.

Robertson, D. (1994). 'The Scope of Facilities Auditing'. *CSM*, vol. 3, no. 10, p. 41.

Rondeau, E.P., Brown, R.K. and Lapides, P.D. (1995). *Facility Management*. John Wiley & Sons.

Tompkins, J.A. (1996). *Facilities Planning* (2nd edition). John Wiley & Sons.

Willis, J. (1992). 'The Inner Game of Facilities Management'. *Facilities*, vol. 10, no. 10, pp. 8–10.

Index

Accounting rate of return, 97, 102
Acid rain, 53
Active heating and cooling systems, 143
Adjusted internal rate of return, 100
Advertising, 233
Affordability, 112–13
After-use monitoring, 47
Agenda 21, 12, 16
Aggregate efficiency, 77, 82, 118
Air conditioning, *see* Heating, ventilation and air
    conditioning (HVAC) systems
Air pollution, 54
Allocative efficiency, 78, 84
Alternatives analysis, 275–6
Aluminium: embodied energy intensity, 155–6, *156*
Ambient permit systems, 72
Ambient standards, 39
Ammonia, *5*
Angelsen, A., 219
Appliances, 147–8
Area-wide environmental impact assessment, 49
Asbestos, 129
Ashworth, A., 219, 221
Asia: population growth, 11
Assembly, 22
Asset management, 249–51
    see also Environmental auditing; Facility management;
        Post occupancy evaluation
Assets management, 279–80
Atmospheric protection, 4
Australia:
    carbon dioxide emissions, 129
    energy auditing, 264
    energy consumption, 169
    energy life cycle analysis, 155
    environmental policies, 55–6
    HVAC systems, 130
    recycling, 158
    telecommuting, 243
Average rate of return, 97, 102
Averting behaviour method, 32

Barrett, P., 272
Bates, G.M., 37–8
Beder, S., 19
Bellinger, W.K., 109–10
Benefit-cost ratio, 97, 100–1, 121
Benefits, 69, 74, 84, 121
    see also Cost-benefit analysis
Bequest values, 28
Biodiversity, 9–11, 20, 52, 54
Biomass, 151, 177–8, 181–2, 190, 191
Bird, R., 222
Bohm, P., 88
Botkin, D., 39
Bowen, B., 221
Bragg, S., 58
Brazil, 182
BREEAM, 201
Broadbent, C., 238
Bromilow, F.J., 220
Broome, J., 110
Brown issues, 54
Browning, W.D., 241

Brownlee, H., 112–13
Brundtland Report, 16, 17, 18–19, 53, 54
    sustainable development definition, xiii, 17
Buffalo Organization for Social and Technological
    Innovation (BOSTI), 239
Building alterations, 233
Building management systems, 266
Building performance evaluation, 253–4
Bullard, C., 151
Busch, J.F., 129, 195
Bush, S., 129

Canada, 133
Capital budgeting, 97, 105–6
Capital costs, 206
Capital energy, *see* Embodied energy
Capital productivity, 222–3, 224, 227
Carbon dioxide, *5*
    global warming, 6–7, 10–11
    policy implementation, 54
    sources, 7–8, *7*, 129–30, 136
Carley, M., 19
Carpets, 132–3
Carson, R., 15, 52
CFCs (chlorofluorocarbons), 8, 54
    developing countries, 41
    insulation, 128–9
    ozone depletion, 15, 52
    regulation, 55, 129
    replacement, 9
Chemical contamination, 52
China, 53
Christie, I., 19
Cities, 53, 54, 108, 134
Clark, B.D., 44
Climate: passive solar design, 139–40
Climate change, 4–9, 10–11
Cogeneration, 132, 171–2, *172*
    district heating and cooling, 172–3, *173*
    efficiency, 190–1
Collective utility, 77, 82, 118
Commission on Sustainable Development (CSD), 16
Commissioning, 22
Common law, 38
Compensation: environmental impact assessment, 47
Computer-aided facility management, 270–2
Concrete: recycling, 158
Conservation groups, 42, 52, 58
Construction cost, 206
Construction monitoring, 47
Contingency planning, 276
Contingent ranking, 30
Contingent valuation, 29–30, 86–7, 87–8, 88
Conventions, 40, 41
Cooling, 173–4
    active systems, 143
    district systems, 172–3, *173*
    passive, 141–2
Corporate strategies, 261–2
Cost analysis, 208
    see also Life-cost analysis
Cost planning, 156, 206
    see also Life-cost planning
Cost saving: energy auditing, 267

Cost-benefit analysis, 61, 62, 74, 82, 95, 116, 201
  fundamentals, 74–5, *76*
  and multi-criteria analysis, 105, 120
  problems, 117–19
  see also Social cost-benefit analysis
Cost-effectiveness analysis, 81, 119–20
Cost-in-use, *see* Life-costs
Costs, 69–70, 74, 84, 126, 203
  and performance, 204
  see also Life-cost studies
Council on Environmental Quality (United States), 45
Cropper, M., 54

Daly, H.E., 26
Dawson, B., 220
Daylighting, 131, 144–5
  and worker productivity, 235, 238
Defensive expenditure technique, 88
Deforestation, 10–11, 68, 128
Democracy: and environmental policies, 55
Denmark, 178
Deposit-refund systems, 71
Depreciation approach, 27–8
Design, 125–6
  and occupancy costs, 234, 240–5, 247
  and sick building syndrome, 237–8
Design optimization, 204
Developing countries:
  aspirations, 53
  CFCs, 41
  hip-pocket principle, 246
Development, 17, 107
Development controls, 35–6
  see also Environmental impact assessment;
    Environmental law; Environmental policies
Devlin, R.A., 57
Diagnostic post occupancy evaluation, 256
Dickson, W.J., 241, 242
Dingle, P., 237, 238
Dioxin, 52, 55
Direct energy, 151–2
  hybrid analysis, 153
  input-output analysis, 153
  process analysis, 153
Direct use values, 28
Discount rate, 75, *76*, 95, 218, 227–8
  capital productivity, 222–3
  determination, 225–7
  further considerations, 223–5
  selection, 220–1
  time preference, 221–2
Discounted cash flow, 75, *76*, 97, 105
Discounted payback, 97, 102
Discounting, 25, 61, 75, 95, 201, 218–19, 227
  externalities and intangibles, 78
  and intergenerational equity, 109–12
  life-cost studies, 203–4, 209–11
  problems, 117, 118
  rationale, 219–21
Distribution, 109
District heating and cooling, 172–3, *173*
Diversification, 20
Dobbs, I.M., 110
Does-response method, 30–1
Domestic resource cost ratio, 97, 101

Earth Summit, 16, 19
Ecolabelling, 133–4
Ecological sustainability, 20–1
Ecologically sustainable development, 53
  facility management, 270, 271–2

Economic appraisal, *see* Social cost-benefit analysis
Economic cost-benefit analysis, 74
Economic efficiency, 69–70
Economic functions, 25
Economic growth, 1, 17, 53
  and entropy, 170–1
  and environment, 19
Economic incentives, *see* Incentives
Economic rate of return, 99
Economic sustainability, 19–20
Economics, 65, 68–9
  and environmental policies, 51
  see also Environmental economics
Ecosystem conservation, 20–1
Ecosystem diversity, 10
Eddington, Sir Arthur, 168
Efficiency, 20, 105, 150
Electrical appliances, 147–8
Electricity, 163, 169
  fuel cells, 189–90
  photovoltaic systems, 147–8
  solar, 183–7
Embodied energy, 128, 150–1, 151–2, 159
  additional validity problems, 155–6
  energy labelling, 200
  extraction methodologies, 152–4
  implications for sustainable buildings, 156–7
Emissions:
  marketable permits, 72
  materials, 128
  standards, 39–40
  taxation, 70–1
  see also Greenhouse gases
Energy, 150, 151, 159, 167, 174
  cogeneration, 132, 171–2, *172*
  and design, 126, 136–48
  laws of thermodynamics, 167–8
  renewable, 176–91
  see also Embodied energy
Energy auditing, 250, 263
  benefits, 267
  future importance, 268
  implementing recommendations, 266–7
  procedures, 264–6
  process, 264
Energy conservation, 163–5, 174
Energy efficiency, 168–70, 174
  benefits, 268
  in buildings, 171
  see also Low energy design
Energy labelling, 198–200
Energy life cycle analysis, 152, 155
Energy policies, 193–4, 201
  energy labelling, 198–200
  incentive schemes, 194–5
  mandatory standards, 196–8
  voluntary standards, 195–6
Energy usage, 117
Enforcement: environmental policies, 56
Enterprise Energy Audit Program (Australia), 264
Entropy, 170–1
Environment, 17, 18
  and economy, 19, 68–9
Environmental accounting, 25–6, 28
Environmental auditing, 58, 261, 268
  corporate strategies, 261–2
  future importance, 268
  process, 262–3
  see also Energy auditing
Environmental conservation, 1
Environmental damage, 121

Environmental economics, 53, 65, 72–3
  environmental management approach, 71–2
  importance, 65–6
  incentive-based approaches, 70–1
  natural environment, 67–8
Environmental goods, 28
  valuation techniques, 29–32
Environmental impact, 127, 134, 228
  alternative approaches, 131
  carpets, 132–3
  cogeneration, 132
  ecolabelling, 133–4
  energy, 129–30
  engineered timber, 132
  fabrics, 133
  HVAC systems and lighting, 130
  materials, 128–9
  natural ventilation and daylighting, 131
  stormwater, 130–1
  visual, 127
  waste management, 133
Environmental impact assessment, 35–6, 44, 49, 120
  ideal conditions, 48
  impacts evaluation, 47
  impacts identification, 46–7
  life-cycle, 49
  management and control of impacts, 47
  scoping, 45–6
  strategic, 48–9
Environmental impact statements, 35, 44–5
Environmental labelling, 133–4
Environmental law, 36, 37, 42, 70
  international, 40–2
  scope, 37–40
Environmental management, 57–8, 71–2
Environmental management systems, 58
Environmental movement, 42, 52, 58
Environmental planning agencies, 38
Environmental policies, 51, 58–9
  approaches, 55–6
  development, 51–3
  issue types, 54
  political process, 54–5
  tradeable property rights, 56–7
Environmental quality, 2
  environmental economics, 65
  and sustainability, 4
Equipment, 232
Equity, 18, 119
  multi-criteria analysis, 105
  natural and human-made capital, 21–2
  see also Intergenerational equity
Ergonomics, 238
Ethos, 275
European Community: ecolabelling, 134
External balance, 20
Externalities, 78–9, 88–9, 118, 119

Fabrics, 133
Facility management, 270–80, 280
  computer-aided, 270–2
  functional plans, 276–80
  measurement and performance standards, 280
  models, 272
  Strategic Facility Plans, 272–3
  strategic planning, 273–6
Feasibility, 95–6, 125, 203
Feedback, 250
  post occupancy evaluation, 258, 259
  strategic planning, 276
Fertilization, 12

Field, B.C., 39–40, 41, 57, 69
Finance costs, 208
Financial analysis, 74
Financial incentives, *see* Incentives
Financial management, 277–8
Financial rate of return, 99
Finland, 173
Flanagan, R., 209, 220, 222
Fletcher Constructions, 133
Fluorescent lights, 145
Forecasting, 103
Forests, 10–11, 68, 128
Form, 126
Freeman, S., 28
Friends of the Earth, 52
Fuel cells, 189–90, 191
Fuel generation:
  biomass, 182
  solar, 182
Function, 126
Functional plans, 276–80
Functional-use analysis, 230–1
Functional-use costs, *see* Occupancy costs
Future generations, *see* Intergenerational equity
Futurity, 17–18

Genetic diversity, 10
Geothermal energy, 187–8, *188*
Germany:
  acid rain, 53
  ecolabelling, 133
  environmental policies, 55
  wind power, 178
Gilchrist, M., 112
Gilpin, A., 45, 48, 49
Global warming, 6–7, 54
GNP, 26
Goodacre, P., 221
Goodin, R., 118
Gouldson, A., 56, 58
Grafton, R.Q., 57
Grain harvest, 12
Green design, 126, 245
  financial incentives, 246–7
  hip-pocket principle, 245–6
Green issues, 54
Green, J., 234, 239
Greenhouse gases, 5–6, *5*
  emissions, 52, 129, 130, 136, 150
  Kyoto Protocol, 16–17, 54, 137
  sources, 7–8, *7*
  see also Carbon dioxide
Greeno, J.L., 261
Greenpeace, 42, 52
Gyourko, J., 108

Habitat diversity, 10
Hafemeister, D., 137
Hatzantonis, S.J., 232, 233, 242
Hawthorn effect, 241–2
HCFCs, 129
Healthy working environments, 237–8
Heat:
  geothermal energy, 187–8, *188*
  thermal comfort control, 138
  thermal mass, 140–1
Heating, 173–4
  active systems, 143
  cogeneration, 132, 171–2, *172*
  district systems, 172–3, *173*
  solar, 182–3

Heating, ventilation and air conditioning (HVAC) systems, 126
  environmental impact, 130
  low energy design, 148
  and sick building syndrome, 238
Hedonic pricing, 31, 87
Herendeen, R., 151
Hip-pocket principle, 245–6
Hoevenagel, R., 29
Housing, 54
Human-made capital, 21–2
Hussen, A.M., 57
Hybrid analysis, 152, 153–4
Hydrogen fuel cells, 189–90
Hydropower, 177–8, 179–80, 191
Hypothetical context bias, 30

Impact matrices: environmental impact assessment, 46
Impatience, 221–2, 224, 227
Incentives, 36, 55, 70–1
  energy policies, 194–5, 196
  environmental management, 57–8
  green design, 246–7
Incidence analysis, 81
Income, 113
India, 178
Indicative post occupancy evaluation, 256
Indirect energy, 28, 153
Indirect use values, 28
Indoor air quality, 128
Inflation, 224, 225, 226
Information bias, 30
Information technology: and facility management, 270–2
ING Bank, Amsterdam, Holland, 235–6
Input-output analysis, 152, 153–4
Insulation:
  incentives, 36
  low energy design, 138–9
  materials, 128–9
  windows, 144
Intangibles, 78–9, 119
Intensity extraction methodologies, 152–4
Interconnectedness, 21
Interest: discount rate, 224–5
Intergenerational equity, 18, 21, 53, 107–8, 114
  affordability and living standards, 112–13
  discount rate, 224, 227
  and discounting, 109–12
  justice and distribution, 108–9
  time preference, 221–2
Intergenerational impartiality, 110
Intergenerational neutrality, 110
Internal rate of return, 75, 97, 98, 99–100, 106
International agreements, 40–1
International Council for Local Environmental Initiatives (ICLEI), 137
International Court of Justice (ICJ), 40–1
International Facilities Management Association, 270
International Framework Convention on Climate Change (IFCCC), 6
International Union for Conservation of Nature and Natural Resources (IUCN), 16
Intragenerational equity, 18
Investigative post occupancy evaluation, 256
Investment, 20
Investment return, 117, 223, 224, 225, 226

Janda, K.B., 129, 195
Japan, 53, 133
Justice, 108–9

Kaldor-Hicks principle, 118
Keller, E., 39
Kula, E., 111–12
Kyoto Protocol, 16–17, 54, 137

Labour productivity, see Worker productivity
Land cost, 206
Land use, 38–9
Landfill, 23, 182
Landscaping, 141
Leary, N.A., 6
Ledgerwood, G., 261–2
LEDs, 145
Liability, 38
Life cycle cost, see Life-costs
Life-cost analysis:
  data collection, 209
  monitoring and management, 209
  occupancy costs, 230–1
  purpose, 208
Life-cost planning, 205, 215
  discounting, 209–11
  purpose, 206, 207
  risk and uncertainty, 212–13
  study period, 211–12
  subjective considerations, 213–15, 216
Life-cost studies, 200, 201, 203–4
  worker productivity, 244
Life-costs, 205, 230
  discounting, 219–20
  types, 206, 208
  see also Occupancy costs
Life-cycle environmental impact assessment, 49
Lighting, 130, 131
  controls, 146
  design, 146–7
  energy-efficient, 145
  worker productivity, 235
  see also Daylighting
Lippiatt, B.C., 238–9, 244
Living standards, 112–13, 114
Lockheed Building 157, California, USA, 131, 236
Low energy design, 126, 136–7, 138, 148
  active heating and cooling systems, 143
  appliances, 147–8
  daylighting and energy-efficient artificial lighting, 144–7
  insulation, 138–9
  landscaping, 141
  natural ventilation, 142–3
  passive cooling, 141–2
  passive solar design, 139–40
  thermal comfort control, 138
  thermal mass, 140–1
  windows, 143–4
Low voltage lighting, 145

McDonalds, 56
McDonough fabrics, 133
Maintenance cost, 208
Maintenance management, 278
Mandatory standards: energy policies, 196–8, 201
Manning, I., 112
Marginal production costs, 70
Marginal willingness to pay, 70
Markandya, A., 30, 220
Market prices, 84–6
Market-based solutions, 55, 58
Marketable pollution permits, 57, 58, 72
Markets, 70
Marshall, H.E., 100, 101

Materials:
    environmental impact, 128–9
    insulation, 138–9
    selection, 22
    thermal mass, 140–1
    see also Recycling
Methane, 5, 8
Methyl chloride, 5
Meyers, S., 4
Mikesell, R.F., 117–18
Minimum acceptable rate of return, 97, 99, 222
Mitigation: environmental impact assessment, 47
Monitoring:
    environmental impact assessment, 47
    life-cost analysis, 209
Montreal Protocol, 9, 54, 129
Multi-attribute decision-analysis, 244
Multi-criteria analysis:
    project selection, 105, 106, 123
    sustainability assessment, 116, 117
    value management, 214
Munasinghe, M., 18, 19, 67

National accounting, 27–8
National Environmental Policy Act (NEPA) (United States),
    44
National Greenhouse Strategy (Australia), 155
Natural capital: conserving, 21–2
Natural environment, 66–7
    role, 67–8
Natural light, see Daylighting
Natural ventilation, 142–3, 238
Net benefit investment ratio, 98, 106
Net benefits method, 244
Net national product, 26, 28
Net present value, 75, 97, 98–9, 105–6, 120, 222
    calculation, 79–80
    intergenerational equity, 109, 110–11
Net social benefit, 84
Network methods: environmental impact assessment, 46
New Zealand, 11, 179
Nijkamp, P., 110–11
Nitrogen, 5, 7
Nitrous oxide, 5, 7, 8
Noise, 54
Non-governmental organizations, 54
Non-renewable resources, 66, 67, 151
Non-use values, 28
Norman, G., 209
Norway, 179
Nuclear fusion, 190
Nuclear testing, 54

Oates, W., 54
Obsolescence, 243
Occupancy costs, 204, 208, 230, 247
    applicability, 239–40
    building design influence, 234
    categories, 231–3
    design constraints, 240–5
    healthy working environments, 237–8
    significance, 230–1
    worker productivity, 234–7, 238–9
Ocean power, 180–1
Öfverholm, I., 220
Oil, 193
Operating costs, 208
Operation monitoring, 47
Operational costs, see Life-costs; Occupancy costs
Operational services management, 278–9
Operational supplies, 233

Opportunity cost method, 31
Opportunity costs, 69–70, 222, 227
Option values, 28
Our Common Future (WCED), see Brundtland Report
Overlays: environmental impact assessment, 46
Ownership cost, 208
Oxygen, 5
Ozone layer, 8–9, 15, 52, 54

Passive cooling, 141–2
Passive solar design, 139–40
Payback period, 97, 101–2
Payment bias, 30
Pearce, D.W., 17–18, 21–2, 25, 54, 72, 90, 220, 221
Pearman, A., 30
Pears, A., 153
Peirson, G., 222
Pennsylvania Power and Light, USA, 235
Performance: and cost, 204
Perkins, F., 85, 87, 88, 99, 101
Photobiological production, 190
Photoelectrochemical cells, 186–7
Photosynthesis, 181–2
Photovoltaics, 147–8, 177–8, 184–6, 185
Pigou, A.C., 55, 220
Planning, 38–9
Plant and equipment, 232
Political pluralism, 54–5
Polluter pays principle, 70–1
Pollution, 12–13
    control, 39–40
    environmental policies, 51
    taxes, 70–1
    tradeable property rights, 57, 58, 72
Pollution offset systems, 72
Population growth, 2, 11–13, 69
Post occupancy evaluation, 250, 253, 254, 259
    benefits, 257–8
    case study, 258–9
    generic approach, 257
    models, 255–6
    role and purpose, 255
Poverty, 12, 53, 113
Power generation emissions, 53, 56–7
Price, C., 109, 118
Primary energy, 171
Primary energy factor, 155
Primary recycling, 157
Process analysis, 152–3
Procurement, 22
Productivity Impact Function, 244
Profit: discount rate, 223, 224, 226
Programmatic environmental impact assessment, 48–9
Project appraisal techniques, 116
    see also Cost-benefit analysis
Project feasibility, 95–6, 125, 203
Project selection, 97, 105–6
    multi-criteria analysis, 105
    risk and uncertainty, 103–4
    techniques, 98–102
Project sustainability, 95–6, 203
Property rights, 56–7, 58, 59, 72
Public goods, 28
Purchase cost, 208

Quality of life, 112, 113, 120
    see also Living standards

Rainforests, 10–11
Ramsey, F.P., 220
Randall, A., 88

Recurrent costs, *see* Life-costs; Occupancy costs
Recycling, 2, 157, 159
  deposit-refund systems, 71
  environmental economics, 68, 69
  environmental policies, 53
  obstacles, 158
  operational supplies, 233
  sustainable construction, 22, 23
Regional impact analysis, 82
Regulation, 55–6, 193–4, 268
Reinhardt, F.L., 58
Renewable energy, 176, 190–1
  background and history, 176–8
  fuel cells, 189–90
  geothermal energy, 187–8, *188*
  nuclear fusion, 190
  photosynthesis, 181–2
  solar electricity, 183–7
  solar heat applications, 182–3
  tidal power, 187
  water, 179–81
  wind power, 178–9
Renewable resources, 66–7, 151
Reno Post Office, Nevada, USA, 234–5, 241–2
Resources:
  entropy, 170
  living standards, 113
Return on capital employed, 97, 102
Rio Conference, 16, 19
Rio Declaration, 16, 54
Risk, 95, 103–4
  discount rate, 223, 224, 226, 227
  life-cost planning, 212–13
  time preference, 221–2
Risk aversion, 21
River power, 179
Roberts, P., 56, 58
Robinson, J.R.W., 219–20
Roethlisberger, F.J., 241, 242
Rosenfeld, A., 137
Rouwendal, J., 110–11
Rowe, D., 238
Ruegg, T., 100, 101
Running costs, *see* Life-costs; Occupancy costs

Salaries, 231, 232, 242, 244, 245
Savings-investment ratio, 97, 100–1
Scale of impact, 21
Schegara, J.D., 6
Schipper, L., 4
Scoping, 45–6
Secondary materials, *see* Recycling
Secondary recycling, 157
Selling cost, 208
Sensitivity analysis, 104, 106, 213, *213*
Serafy, S.E., 27–8
Service, B., 231
Shadow pricing, 61–2, 85
Shechter, M., 28
Sick building syndrome, 128, 237–8
Silent Spring (Carson), 15, 52
Simulation models: environmental impact assessment, 47
Sinden, J.A., 84, 87, 92
Site planning and organization, 22
Social benefits, *see* Benefits
Social cost-benefit analysis, 74, 82, 92, 116–17
  advantages and disadvantages, 80–1
  aggregate efficiency, 77
  allocative efficiency, 78, 84
  cost-effectiveness analysis, 81
  externalities and intangibles, 78–9, 88–9

incidence analysis, 81
methodology, 79–80
problems, 118
purpose, 75–7
regional impact analysis, 82
social impact analysis, 82
sustainability constraint, 89–92
valuation with market prices, 84–6
valuation without market prices, 86–8
Social costs, *see* Costs
Social impact analysis, 82
Social opportunity cost of capital, 222
Social time preference, 221–2, 224, 227
Solar access, 139–40
Solar energy conversion, 177
  direct, 182–7
  indirect, 178–82
Solar heat gains, 143–4
Solar heating, 143
Solar water heating, 147
Space conditioning, *see* Heating, ventilation and air
    conditioning (HVAC) systems
Space heating, *see* Heating
Space management, 277
Spain, 178
Species diversity, 10
Species extinction, 11, 52
Stakeholders, 274
Standards of living, 112–13, 114
Stated preference method, 30
Stone, P.A., 220, 221
Stormwater, 62, 130–1
Strategic bias, 30
Strategic Environmental Assessment, 48–9
Strategic Facility Plans, 272–3
Strategic planning, xv, 273–4
  alternatives analysis, 275–6
  contingency planning, 276
  environmental/contextual factors, 274–5
  feedback, 276
  measurement and performance standards, 280
  stakeholders, 274
  Strategic Facility Plans, 272–3
Strategy, 275
Stratospheric ozone layer, 8–9, 15, 52, 54
Study period: life-cost planning, 211–12
Styrene-butadiene rubber, 132–3
Subsidies, 71
Sulphur gases, *5*
Sustainability, 3
  measurement, 116–17
  multi-criteria analysis, 105
  principles, 19–21
  see also Project sustainability
Sustainability constraint, 89–92
Sustainability index, 120–1, *122*, 123
Sustainable construction, 22–3
Sustainable development, 1, 15, 23
  Brundtland Report, 18–19
  conceptual framework, *20*
  definitions, xiii, 17–18
  and economic growth, 19
  historical context, 16–17, 53
  intergenerational equity, 107–8
Sustainable income, 26–7
Sweden, 181
SWOT analysis, 275

Tactical planning, 276–7, 280
Taxation:
  discount rate, 223, 224, 225, 226

environmental policies, 56, 70–1
Taylor, D.C., 120
Technology, 126
Technology standards, 40
Telecommuting, 243
Tendering, 22
Terotechnology, *see* Life-costs
Thampapillai, D.J., 28, 67, 84, 87, 92
Thermal comfort control, 138
Thermal mass, 140–1
Thermodynamic conversion, 183–4
Thermodynamics, 167–8
    energy efficiency, 169, 171
    entropy, 170
Tidal power, 187
Tietenberg, T.H., 57
Timber:
    embodied energy intensity, *154*, 155–6, *156*
    engineered, 132
Time horizon, *see* Study period
Time preference, 221–2, 224, 227
Total cost approach, 205
Total energy, 121
Toxic gases, 132–3
Toxic waste, 55
Tradeable property rights, 56–7, 58, 72
Traffic congestion, 54
Travel cost method, 32
Treaties, 40, 41
Trees, 141
Treloar, G.J., 151, 153
Tucker, S.N., 221
Turner, R.K., 72

Ultimate cost, *see* Life-costs
Uncertainty, 103–4
    discount rate, 224, 227
    life-cost planning, 212–13
    time preference, 221–2
United Kingdom:
    BREEAM, 201
    wind power, 179
United Nations:
    Commission on Sustainable Development, 16
    international agreements, 40
    International Court of Justice, 40–1
    see also World Commission on Environment and
        Development
United Nations Conference on Environment and
    Development (UNCED), 4
    Agenda 21, 12, 16
    biological diversity, 9, 10
    Earth Summit, 16
United Nations Conference on the Human Environment,
    Stockholm, 16
United Nations Environment Programme (UNEP), 11, 16
United Nations Framework Convention on Climate Change,
    9
United States:
    carbon emissions, 129–30
    Department for Housing and Urban Development, 49
    economic incentives, 55
    emission taxes, 57
    energy consumption, 194
    environmental impact assessment, 44, 45
    environmental policies, 55
    wind power, 178
Urban heat islands, 130

Urien, R., 220
Use values, 28
User cost approach, 27–8

Valuation, 92
    with market prices, 84–6
    without market prices, 86–8
Value management, 214
Value for money, 120, 203, 204
Value for money index, 214–15, *216*
Van Pelt, M., 105
Ventilation, *see* Heating, ventilation and air conditioning
    (HVAC) systems; Natural ventilation
Vernacular architecture, 137
Vienna Convention for the Protection of the Ozone Layer,
    9
Visual impact, 127
Voluntary standards: energy policies, 195–6

Wal-Mart, Kansas, USA, 235
Ward, R., 244
Waste disposal, 54, 68, 128, 233
Waste minimization, 22–3, 53, 56, 133
Water:
    hydropower, 177–8, 179–80, 191
    ocean power, 180–1
    river power, 179
    tidal power, 187
    see also Stormwater
Water heating, 147, 182–3
Water pollution, 54
Wave energy, 180–1
Weber, S.F., 238–9, 244
Weighted cost of capital, 222, 223, 224, 225, 226, 227
Welfare enhancement, 116, 117
West Bend Mutual Insurance, Wisconsin, USA, 236–7
Whalley, J., 54
Whole-of-life methodologies, xiv
Wigle, R., 54
Wilderness preservation, 52, 54
Wilke, S., 238
Willingness to accept, 29–30
Willingness to pay, 29–30, 86–7, 87–8, 88
Willis, K.G., 224
Wind power, 177–9, 191
Windheim, Lee, 236
Windows, 143–4
Winters, L.A., 54
Worker productivity:
    case studies, 234–7
    computer-aided facility management, 271
    design influences, 238–9
    design knowledge, 242–3
    energy auditing, 267
    Hawthorn effect, 241–2
    healthy working environments, 237–8
    measuring changes, 244–5
Working environment, 237–8, 267
Workload, 245
World Commission on Environment and Development
    (WCED), 16
    see also Brundtland Report
World Conservation Strategy (WCS), 16
World Wildlife Fund, 16, 42
Wright, M., 98

Zarsky, L., 20–1
Zero population growth, 69